They Don't Make Them Today Like They Used To

1st Edition

Mark A. Lester

They Don't Make Them Today Like They Used To

By Mark Lester

Copyright 2021 by Mark A. Lester

Self-published through LuLu

All rights reserved. No part of this publication may be reproduced, distributed, or transmitted in any form or by any means, including photocopying, recording, or other electronic or mechanical methods, without the prior written permission of the publisher, except in the case of brief quotations embodied in critical reviews and certain other noncommercial uses permitted by copyright law.

Table of Contents

Introduction

The Early Years	Page 1
Kenny & Joe Join the Work Force	Page 7
Memories & Stories from Which Legends are Created	Page 26
The Menagerie	Page 36
Pugilism	Page 45
Heroes	Page 58
Ironworker for a Day	Page 61
Kenny Obituary	Page 66
Uncle Joe	Page 68
Scotland Cedar Springs Connection	Page 72
Canadian Black Bear Hunt	Page 76
The Curse	Page 81
Drib	Page 87
Final Thoughts	Page 92

"They Don't Make Them Like They Used To"

Introduction

My mother was blessed with four brothers, three older and one younger. She was their only sister. They grew up in a small community situated in the hills of southern Indiana, during challenging times which included a severe extended economic depression and a world war. My mom would tell that during her childhood, "they were poor but didn't know it as everyone else was in the same situation". Hard tough times breed hard tough individuals which is true in the case of my mother and her brothers. In writing this journal, my goal is to capture and share the life experiences of my mom and her brothers with the major focus on the exploits of her two oldest brothers, Kenny and Joe Asdell. These stories have been shared with me throughout my life by many different individuals. I feel I have a choice, document these stories so future generations can learn about the lives of some "colorful" ancestors or let the stories "die on the vine". I prefer the first choice, document them for posterity. Inside are stories of people who were shaped by their times thus forming a generation that may never be duplicated.

My Uncles, Kenny and Joe, were determined, adventurous, fearless, and strong individuals, traits which led them on many adventures and into some perilous situations. These traits also contributed to behaviors that would be deemed "socially unacceptable" and, at times downright abusive". Just so there are no illusions that these guys were "angels" or candidates for "sainthood". My intention is not to glorify or assign hero status but rather to share the exploits of two tough men who endured tough times and persevered. If I have learned nothing else in life's journey, it is that no human being is totally "good". There were times when my uncles "moral compass" would become askew.

I often equate my Uncle Kenny with John Wayne as they did share some similarities. Both were physically large men, both were bold, both were personable. However, John Wayne was "make believe", he made his livelihood filling theater seats by staring in macho movie roles. Kenny Asdell, however, was "the real thing". He lived the life that John Wayne portrayed in his acting roles. He was never a "pretender" and neither was his younger brother, Joe.

The Early Years

Kenneth Elwood Asdell was born July, 4, 1927, at the homeplace of his Grandparents, Joseph & Cora Kern, located just north of Scotland, Indiana. The very same homestead where his mother, Sophia (Kern) Asdell had been born in 1908. July 4th was definitely a very fitting birthdate for Kenny. On August 25, 1929, Kenny was joined by a baby brother, Davie Joe Asdell. Joe was born in New Castle, Indiana, where the family had relocated when their father, Leland "Drib" Asdell, found employment at the Chrysler plant.

Kenny Asdell
1939-1940 Scotland School

The family later moved back to Scotland, Indiana, where Kenny and Joe were joined by a brother, Benoni W. Asdell, on March 1st, 1932. It was at this house where Kenny allegedly participated in his first fight. Billy Ed Hostetler, longtime family friend and preacher, shared at the funeral of my grandfather, Leland "Drib" Asdell how he was first introduced to Kenny and his dad by getting into a fight with Kenny. Seems, Billy Ed was accompanying his Grandfather Hostetler to Scotland whereupon they stopped to visit with my grandpa "Drib" Asdell. Kenny was next to his dad on a tricycle. Billy Ed said he was barefoot and Kenny proceeded to run over his foot with the tricycle. Billy Ed told Kenny he better not do that again or he was going to get it. Well, Kenny did it again and the fight was on. That was Billy Ed's first introduction to Kenny Asdell as well as the first Kenny Asdell fight on record. In time, Billy Ed and Kenny became very good friends

The next move was to Doans, Indiana, where my mother, Doris June Asdell, was born on November 30, 1934. Their home at Doans was situated on a large hill south of the Doans Store. Uncle Joe told me that the Doans house had no insulation, no heat in the upstairs, and large gaps between the siding which allowed snow to enter the bedroom where he and Kenny shared a bedroom. It was at Doans, where Kenny fell from a large hickory tree and knocked himself unconscious. In the panic to get Kenny to the doctor, my grandmother left my mom lying on a blanket in the yard. Not sure how long it took her to realize she was missing a child?

My grandpa Drib Asdell supported his family in various occupations, including teaching school. However, at Doans he worked as an independent trucker. His Chevrolet truck was equipped with a "mechanical trip" dump body in which he hauled crushed limestone. I remember Uncle Joe talking about having to hang off the dump body to get it to come back down when it was raised on an incline. Depending upon the distance to the quarry from his house where he was hauling, my grandpa would often spend the week at the site. He worked with his old friend Eddie Hostetler, uncle to Billy Ed. My grandpa told me depending upon the weather, they would sleep either in back of the truck or underneath it. Uncle Joe told me that his dad kept spare truck tires and tubes as flats were commonplace back in the day of natural rubber tires with inner tubes. He said his dad would spend weekends repairing tires in preparation for the upcoming work week. Which he could do only after he retrieved the tires that Joe and Kenny had rolled down the hill. Joe and Kenny took turns trying to ride inside the tires while they rolled down the hill.

The next move took them from Doans back to Scotland into the home where my Grandparents would spend the rest of their life. Uncle Joe told me that his Grandpa Dave Asdell used his team and wagon to move them back to Scotland from Doans. It was in this home where the last Asdell sibling was born, Billie Nelson Asdell made his appearance on Christmas Eve 1941. Joe told me that he was helping his dad assemble a doll house for my mother, June, for Christmas when his mother went into labor. Joe, Kenny, and Ben were sent to the home of their neighbor, Amy Isenogle, to spend the night in her "straw tick guest bed" while their mother gave birth to their new little brother.

Many of the old homes in Scotland, Indiana, including the home of my grandparents, had no foundation. Instead, they had large beams which rested on flat rocks keeping them up off the ground. According to my mom, when Kenny was a small boy, he wandered up to Whitaker's Store and procured bananas which he charged to my grandparents' account. He stuffed the bananas down his bib overalls and crawled under a building to eat them. However, his plan was foiled when he became stuck due to the bulging bananas inside his britches. I do not recall how he got loose.

On another occasion, he attempted to catch a skunk which he mistook for a kitten. He learned a very unpleasant life lesson concerning the difference between a skunk and a kitten.

Once, when Joe was a boy, he was helping demolish a chicken house in town when an escaping rat ran up his leg underneath his bib overalls. He hastily discarded the bib overalls in the middle of town creating a bit of a spectacle.

Joe told me that their family acquired a Jersey milk cow from Joe's and Kenny's Grandpa Dave Asdell which they kept pastured north of Scotland. Twice a day, it was the job of Kenny and himself to retrieve the cow and bring it to their house for milking. He said one day they got the bright idea that they would take turns riding the cow. Kenny mounted the cow first. He sat on the back haunches of the cow and Joe twisted its tail around to Kenny for a hand hold. The plan was for Joe to lead the cow while Kenny rode. Joe said, "I got to thinking about all the dirty tricks Kenny had played on me and decided it was time for revenge so I let go of the lead rope. The cow was nervous and spooky by nature and even more so with Kenny on her back. She took off as fast as she could go heading for the barn at their house. I was laughing so hard I couldn't keep up. Kenny was swearing and calling me every name that came to mind. As I came around the corner of the alley, I observed the cow had indeed made it to the barn, but poor Kenny had not cleared the door. He was sprawled out right in front of the barn door". Joe said, you know, I almost felt bad.

Joe told me that Kenny would get him, Joe, to push him in their wagon. When Joe got the wagon going fast Kenny would drop the tongue into the dirt causing the wagon to screech to a halt and Joe to fall down and skin up his knees. Joe said that he would get mad and refuse to push Kenny anymore but Kenny would plead with him promising to never do that again. Sure enough, eventually Kenny would once again drop the tongue and Joe would once again get skinned up knees. Joe told me that is the memory that went through his mind as he saw Kenny laying in the dirt. Kenny suffered no long-term effects from his rodeo experience.

Scotland, Indiana. 1936. This would have been about the time frame when my grandparents purchased their house on Canal Street and moved the family back to Scotland from Doans.

Dave Whitaker standing in front of his store holding onto flag pole. My Grandpa Drib ran this store when I was a little kid. My Uncle Ben worked at the store in his youth and I believe my Uncle Joe did as well.

My mother told me that her brothers would smoke cigarettes and if she was with them, they would use their cigarettes to put burn marks in her clothes so she wouldn't tell on them. If she did, they could point to the burn marks and accuse her of smoking as well. I don't remember which brother or brothers specifically did this, I know Kenny never smoked, in his adult life anyway. She said when he was a kid, her brother Joe would put firecrackers inside her dolls and blow their heads off. I guess it was challenging being the only girl.

My Grandpa Drib Asdell told me that they had a fox terrier dog when the kids were little by the name of Pete. He said there was an old man, they called Old Man Small, who had a hound that would follow him to town every day.

My Granddad said Joe didn't talk real plain when he was a little boy. Joe would stand and watch for Old Man Small and his hound and when he appeared, Joe would sic Pete on Old Man Small's hound dog. Actually, Joe would say "hic'em Pete". One day Joe took off exploring on his own and when my Grandpa Drib found him and asked how he thought he would find his way home, Joe looked down at Pete and said "put my nose to the ground and smell my way home, I guess".

The old Whitaker store building now owned by Brent Dillman, owner of Snapper Sales. Picture taken in 2019.

Joe told me that when he was in high school, he made a deal with the basketball coach to quit smoking during basketball season and in return the coach would overlook his smoking the rest of the year. He said, sometimes he would smoke with the basketball coach during the non-season.

Kenny left school his senior year and enlisted in the Navy to join the war effort. I am not sure of all his assignments; I do know he did a stint on the east coast performing Nazi submarine watch duties in the area where the Navy was training carrier pilots. He spent some time at sea but I am not knowledgeable on his assignment nor the duties he performed there either. I remember my Grandpa Drib telling me ship duty was where Kenny developed his desire and honed his abilities as a fighter. When there were disagreements, they were settled in a ring with boxing gloves. Apparently, Kenny had plenty of opportunities and experience in the boxing ring.

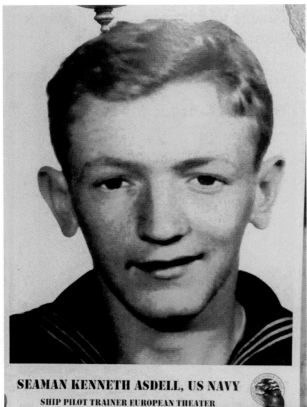

Kenny was left-handed as I am and my grandfather's sister Peg Asdell was as well. I have heard from several sources that Kenny was endowed with a devastating left punch. I read that Teddy Roosevelt was the driving force behind the Navy's implementation of boxing on its vessels. Teddy was a big fan of boxing.

A couple of "lefties".

Mark and Peg

Scotland Elementary & High School
Scotland, Indiana
1912 - 1955

Scotland School. Which was attended by my Grandfather Leland "Drib" Asdell, Great Aunt Margaret "Peg" Asdell, Uncles Kenny Asdell, Joe Asdell, Ben Asdell, my mom June Asdell, and Uncle Bill Asdell. Bill attended up through 8th grade when it was subsequently closed amid the push for school consolidation.

My Grandpa Drib Asdell taught at this school as well.

Kenny & Joe Join the Work Force

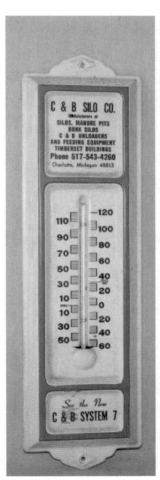

Upon being discharged from the U.S. Navy and returning to Scotland, Indiana, Kenny hired on with C & B Silo in Michigan. C & B Silo was founded by two Navy buddies from World War II, John Cooper from Grand Rapids, Michigan and Ralph Baird from Scotland, Indiana. Finding workers in Michigan was difficult due to the increase in demand for whole goods and ensuing increase in manufacturing which required a huge influx of workers. Ralph recruited workers from the Scotland, Indiana area where labor wasn't in such high demand. Kenny signed up to erect silos in Michigan taking brother Joe with him. They erected silos over most of Michigan except the very southwest corner where C & B had a dealer who provided his own crew. They traveled to Michigan prior to construction of the interstate highway system causing travel to be slower and more dangerous. Kenny engaged in a "cat and mouse" game with the police force in Three Rivers, Michigan, whenever he passed though their city. Apparently, Kenny had outrun a patrol car on one of his trips through Three Rivers causing the police to keep an eye out for his vehicle. They never caught up with him.

Kenny and Joe did not always work together. C & B Silo would generally send two employees to the farm to erect the silo and the farmer was required to provide two laborers. If the farmer could or would not participate then C & B would send extra help. The crew would bring the silo staves and hoops on a trailer. Joe told me the company truck was a Model A Ford flat bed which they used to tow a trailer loaded with concrete silo staves, hoops, scaffolding, white wash, and tools. He said the truck was equipped with mechanical brakes and always overloaded. "Lord have mercy" if you ever had to make an emergency stop. Erecting a silo was very strenuous and physically demanding work which built a muscular physique. The crew lodged with the farmer for whom they were building the silo. At one of the farms where Kenny and his helper erected a silo, Kenny's helper jokingly told the farmer's wife that Kenny would eat a dozen eggs every morning for breakfast. The farm couple had no children and the wife didn't realize Kenny's helper was joking so she fried Kenny a dozen eggs for breakfast every morning he was there. Rather than hurt her feelings and waste the eggs, he ate them. He told me it took him a year before he could "look a pullet straight in the eye". Joe told me they worked out a ploy to pretend one of the crew was having his birthday so the farm wife would bake them a birthday cake.

Joe erected silos in the Frankenmuth, Michigan area for farmers of German descent. He said the farmers he lodged with did not drink water, only beer. Home brewed German beer at that. If you wanted water when you were working out in the hot sun, you were out of luck. They gave you beer. If you wanted water, you had to get it for yourself.

Joe also erected silos in the Grayling area where the ground is very sandy, quite unlike the heavy red clay of southern Indiana. The locals told him you had to set your dog on a board so he could bark othwerwise he would get a mouth full of sand. Joe aslo saw his first "stone boat" while in Michigan which was totally foreign to him. When he inquired as to what they used it for, he was told there was not enough stone in the Michigan fields so they used the stone boat to haul some in. Which is totally the opposite of the truth, Michigan fields tend to be very rocky and a stone boat was used to haul rocks out of the field.

The silo erection crew was required to collect the money for the silo once it was completed. Normally they had no trouble but at one farm the farmer refused to pay Kenny. Joe was with him when it happened. The farmer was trying to find fault with the installation to avoid paying. Kenny told him, if he wasn't going to pay for it then they would have to disassemble it and take it back with them. Kenny started out the door and headed toward the silo with Joe following. Before they actually began the disassembly, the farmer appeared with the check. After they departed the farm, Joe asked Kenny if he was serious about tearing the silo down and Kenny said, "I don't know, I haven't gotten that far yet." You must keep things in context, when Kenny and Joe first went to Michigan to erect silos, Kenny was ninteen years old and Joe was seventeen, basically kids.

Match cover advertising Ravenna Livestock Sales which was owned by J. Paul Herman. Kenny erected silos for J. Paul Herman. Herman ran a large cattle feedlot, a Chevrolet car dealership, and a livestock auction. In the 1970's, my Mom drove schoolbus with Paul Herman's daughter, Colette Lund, at Cedar Springs Schools in Cedar Springs, Michigan.

Kenny had acquired an appetite for fighting while he was in the Navy and he had plenty of opportunities while he was in Michigan. He attended the Golden Gloves boxing championships in Grand Rapids, Michigan (which was where C & B Silo was headquartered in those days and where the concrete silo staves were manufacturered).

After the matches were over Kenny mixed it up with some of the contestants behind the auditorium. He told Joe later that "they could hit you a lot of times but not hard enough to hurt you". Kenny and Joe attended a wrestling event of some type where the promoters were seeking challengers from the audience.

Kenny told Joe that he might just take up the challenge. Joe told him to go ahead but he thought he might just watch. Not sure if Kenny actually participated.

On one of their trips, they were challenged to a fight by some locals at a Grand Rapids restaurant. Kenny wasn't about to let that opportunity "slip through his fingers" so he readily accepted their challenge. There were three local guys versus Kenny, Joe and a Michigan native by the name of Al Shook who worked for C & B Silo as well. Joe said Al Shook was a short guy, quick as a cat, and loved to fight. The local boys had Kenny and Joe follow them out into the country where they drove to a secluded gravel pit. Joe said he was concerned they would come out with guns or knives, but they didn't. He said those young guys had no idea what they were up against. Joe said when the fighting ended, two of the local boys were unconscious and the guy Joe had been fighting was hiding under his car.

Kenny erected a grain bin at the Sand Lake Elevator in Sand Lake, Michigan. The grain bin consisted of stacked up silos. Kenny and his crew got into a fracas in Sand Lake. It was in Sand Lake where a man's eyeball popped out of his head as a result of the fight.

C & B silo grain bins at Sand Lake Elevator in Sand Lake, Michigan, erected by Kenny Asdell and his crew. Kenny and Joe's crews were combined later when they erected a bean storage bin at Quincy, Michigan. At Quincy they added another "layer" of silos. That bin was three high. There was, at one time, a picture circulating of the bin at Quincy after it was completed with Joe standing on the top of it. I think the total height was around 60 ft.

*note*There is a story further back in this publication about C & B silo and the time when Joe fell and landed on a paint bucket while erecting a silo in the Fremont/Whitehall, Michigan area.

There was a local man working on Uncle Joe's crew who lacked motivation. Joe wanted to let him go but they were short handed and this guy was slightly better than nobody. Joe said he had to "motivate" this guy every day. Joe said one day "out of the blue", the guy said "if we are going to work like horses, why don't we act like horses and stand around with our "dicks" hanging out". I don't think that was an option, however. When Joe found out they were to hook up with Kenny's crew at Quincy to build the bean storage bins he told this guy, "You are about to meet up with my older brother and your whole life is going to change". Joe said he was looking forward to see Kenny's reaction to this guy's work habits. However, when they combined crews at Quincy, nothing happened between Kenny and this subpar employee. Joe said I couldn't figure that out as it was doubtful this man upped his game so he said I started paying more attention to what was going on. I came to notice that this guy stayed far away from Kenny.

Whatever side of the grain bins Kenny was working on, this guy would be on the opposite. Joe said this worked for a while but before they completed the bins, Kenny got wise to him and the guy did not have a good day. I think Kenny ran him off.

The trip between southern Indiana and Michigan could be very harrowing back in those days and there were several close calls. One story that my mom shared concerned Kenny encountering an abandoned flatbed semi-trailer in northern Indiana that was partially obstructing the roadway. Kenny was traveling at a high rate of speed in the dark when he came upon the unlit trailer which was not equipped with reflectors. He had very little time to react before colliding with the trailer so he laid down across the seat just prior to the car plowing under the trailer. The quick action on Kenny's part prevented him from being decapitated as the top of his automobile was sheared off.

Kenny later constructed some grain bins in Canada. Joe didn't join him and I don't know the exact location in Canada where he worked.

There were several Scotland, Indiana, natives that traveled to Michigan and worked for C & B Silo over the years. I once interviewed Uncle Kenny about this very subject and he shared with me the names he remembered which included: Bobby Sparks, John Achors, Jim Achors, Jack Inman, George Bailey, Darrell Morgan, and more. Jim Achors married a Michigan girl and started his own silo Company; Tri-State Silo located in Eaton Rapids, Michigan. Jim Achors is no longer with us but the company he founded is still in business, however Tri-State Silo is now headquartered in Riverdale, Michigan. Darrell Morgan, whom Kenny referred to as a "one man basketball team" due to his level of play on the Scotland Scotties basketball team, also married a Michigan girl and became a Michigan native. Darrell worked for the Michigan Department of Transportation at the Mason Maintenance Garage (south of Lansing on US127). I worked for the Michigan Department of Transportation as well, only on the west side of the state in the Kalamazoo area. Darryl and I once compared notes at the Scotland Festival, while we never worked together, we did know some of the same people. Darrell retired about the same time I hired in, in the early 80's.

Another Scotland native who relocated to Michigan was (Gilbert) Vernon Mortland. Vernon had been a school teacher in the Scotland area; he taught both my Uncle Kenny and Uncle Ben. He was drafted into the Army during World War II and while stationed in California with a man from Michigan was introduced to that man's sister, Dorcas, whom he began courting, which led to them getting married. Dorcas Mortland's parents owned a business in Fremont, Michigan, which she and Vern eventually took over. Docas was very active in the community. She was on the Chamber of Commerce in Fremont and the on the board of the local bank. To this day the Fremont Chamber of Commerce awards the 'Dorcas Mortland Award" on an annual basis to a deserving community member.

There are several stories about Kenny and Joe's activities while they were back in Scotland on break from C & B Silo. Joe shared a story about a brawl in which they participated at a small café in Crane, Indiana. Servicemen, Marines primarily, were being processed through the Naval Base at Crane on their return to civilian life. Some of these guys would go into town to "let out some steam". One night some Marines ran afoul of Kenny at this café and the fight was on. I do not know how many Marines participated but Joe told me that he backed himself up next to a cigarette machine which was located by the entrance door and, per his words, "as

Kenny "busted heads" I tossed them out into the parking lot. He said when we left the café there was a parking lot full of "out of commission" Marines.

A story told by Grandpa took place at the same café. It seems my Uncle Joe was entering the café and noticed a Marine who was wrapping his belt around his fist with the buckle up as he approached the door. This was referred to as "poor man's brass knuckles". Kenny was following the Marine and it appeared they were heading outside to square off. Joe realized his brother was about to get "bushwhacked" so he "cold cocked" the Marine as he stepped through the door saving Kenny from an "unpleasant surprise" and which could result in serious injury.

The next adventure for the Asdell brothers was in the Tipton, Missouri, area where they found employment installing a new power line in the foothills of the Ozarks. It was the first electrical power in that area. Tipton is on the edge of the "Land of the Ozarks". The area was very rugged and the ground was mostly rock. They had no power augers or bucket trucks.

The postholes were hand augured, although my Uncle Joe told me most poles were 'dynamited in"
as it was virtually impossible to dig a posthole. The crew had to climb every pole, using **spikes similar to those in the picture** which was an art form in and of itself. Sink them too deep and they are hard to pull out which takes a toll on your legs, sink them too shallow and you risk slipping. They would drape the power cable over their shoulder, climb the pole, and drape the cable over the crosstie. They repeated this on every pole, of course, there is more than one cable on each pole. I found a set of the spikes they used in the bunkhouse at my grandparents' home. The leather harness had deteriorated so they weren't useable but might have been salvageable. Uncle Joe told me the crew had a team comprising of a small donkey and small pony that pulled a trailer on which the cable reel was mounted and that's how they laid out the cable. No dozer in this operation. The team had been "broke" to verbal commands which meant they had no reins or lines nor did they need any. Uncle Joe told me that this team was more valuable to the company than any of its human employees.

The crew stayed in a linesman shack while working seven days a week. Unbeknownst to both Joe and Kenny, their lives were about to be forever changed. On one cold rainy Sunday, the crew was holed up in a linesman shack waiting for the weather to clear so they could go to work. The shack was heated by a small woodstove. The fire in the stove had dwindled so one of the guys poured some coal oil on the embers to reignite it. However, the flames roared up violently and followed the coal oil back to the can. The guy holding the can tossed it to avoid getting burnt, with it landing on Uncle Joe and igniting him. Joe ran out of the linesman shack engulfed in flames. Joe told me that he ran because they stored their dynamite in that shack and he was trying to get away from it. Somebody threw him down and extinguished the flames but not before Joe was severely burned. Kenny got him in the car and headed to a local doctor's office.

Kenny said Joe walked into the Doctor's office under his own power even though his legs were seriously burned and his body fat hung off him like tallow, at least what little body fat he had at that time. The doctor could not do much except give him a shot of morphine and send them to the hospital. Kenny transported his brother to a hospital where the doctors gave him zero chance of survival. They basically kept him comfortable in the belief he would not survive very long. After a day or so they began to realize that maybe he was stronger than they originally thought so they started treating his wounds which consisted of the very painful process of scraping off burned dead skin, sanitizing the wounds, and dressing them. The doctors did not believe they could save his legs and talked of amputation but Joe was having none of that. Kenny got in contact with their parents (my grandparents) and told them if they wanted to see Joe alive, they better get out there. Joe was in a ward with other patients and his dad (my grandpa) did not even recognize him. My Grandpa told me that he and my grandma stayed in a boarding house while they were out there and on the first night he got out of bed and went outside to smoke. He said he was about to go "berserk" from worry. Finally, he thought to himself, "you old fool, you need to get it together. Your family needs you to be strong and supportive.
My mom told me Kenny spent all his money so he could be with his brother, even going so far as selling his winter coat. Joe endured months of painful treatment and skin grafts but he slowly healed and recuperated. He reached the point where it was time for plastic surgery and the doctors gave him two options for hospitals.

One of those hospitals was in Indianapolis so that was a "no brainer" for Joe. He was going back to Indiana and not in a hearse. Although, Jenkins Funeral Home in Bloomfield, Indiana, did provide the transportation.

Jimmy Jenkins told me himself, that he, Jimmy, and his brother Blue took the Jenkins Funeral Home ambulance to Missouri and picked up Joe and bought him to Indianapolis. ***This was back in the days when the local funeral homes provided ambulance services*** I think my grandmother and one other person accompanied them. Joe was still in pain and had to be partially sedated during the trip. Jenkins brothers would not accept any payment for transporting Uncle Joe.

When Joe returned home from the hospital he was confined to a wheelchair. The doctors never expected him to walk again. But walk he did, with the help of his baby brother Bill, he used the wheelchair as a walker and slowly regained his ability to walk. He became his own physical therapist, with Bill's help of course. The doctors also told him he could be addicted to morphine, but that didn't come true either. Doctors didn't expect him to live, but he did. Doctors figured they would have to amputate his legs, but he wouldn't let them. Doctors didn't expect him to walk again, but he did. Doctors expected him to be addicted to morphine and pain killers, but he wasn't. Guess the Doctors didn't know Joe Asdell very well.

I was told the company for which Joe & Kenny were employed refused to pay for Joe's care, nor could he get any workman's compensation, as the company claimed "he wasn't actually working due to the fact it occurred on a Sunday" so as a result my Grandparents paid the medical bills. Kenny told me that my grandparents hired a lawyer in Bloomfield who sued the company to get compensation. My grandparents were led to believe, by this attorney, that they lost the case. However, Kenny told me the lawyer's son told Kenny years later, after this man's father had passed away, that his dad did get the money to compensate my grandparents but he kept it. To my knowledge, my grandparents never were told.

I don't know exactly what Kenny was up to while Joe recuperated, this may have been the time period during which he was building grain bins in Canada. I do know that in 1953 he joined the Ironworkers Union, Local 22, in Indianapolis, Indiana and began building bridges, buildings, and other steel structures.

Somewhere in this same time frame he purchased a 1948 Autocar semi-tractor and began "trucking". I would assume he did these two jobs concurrently, although I do know he had others that drove for him, including his brother Ben. Trucks in those days were pretty primitive affairs compared to those of today. This Autocar was a single axle tractor with a sleeper cab (a very narrow one at that, not at all like those found on today's trucks), was equipped with a four-cylinder Cummins diesel engine, no power steering for which the O.E.M. compensated by installing an extra-large steering wheel, and it had what was commonly referred to as a "Brownie" transmission which consisted of a main gearbox with an auxiliary. As I recall, Uncle Joe told me this truck had a five-speed main box and a three-speed auxiliary, fifteen speeds total.

These things were a real treat to shift, especially on grades (uphill or downhill). I have personally driven modern trucks with 10 and 13 speed transmissions that have multiple ranges but all consolidated in the main gear box and use an air shift for range changes. The big difference being the newer trucks have one gear shifter with an air switch rather than two or three shifters requiring three or more hands when changing ranges during the shift.

Kenny's 1948 Autocar

That old truck holds a "special place in my heart". I spoke at Kenny's memorial service and shared the impact that my Uncle Kenny and his Autocar semi-tractor played on my life and I have included that story in this book. In a "nutshell", my dad died when I was just a few months shy of five years old, leaving my mom a widow with three little kids at 23 years old. There was a lot of sadness and not much to look forward too, except for one thing, and that was my Uncle Kenny who would regularly appear at our home in his Autocar with a bag of groceries for my mom and would take me for a truck ride. For just a short time, this little boy would be on top of the world. I still remember to this day, climbing up into that huge truck and sitting next to my "larger than life" uncle Kenny. It has been over sixty years and that memory still "moistens my eyes". Everyone in our family has memories and stories about Kenny and that old truck. Uncle Ben "trucked" for Kenny and still, to this day, talks about the difficulty in shifting the 'Brownie" transmission. My Uncle Joe talks about climbing the mountain grade to Monteagle, Tennessee, where the truckers would congregate. This was prior to completion of the interstate highway system. The Autocar was underpowered and probably overloaded so it was a slow trip over the mountains. Joe told me they were climbing to Monteagle at a "snails' pace" when the truck started gaining speed. Looking in the mirror they saw the front end of another semi as it was pushing them up the grade.

On a different trip up Monteagle, Joe said the roadway was slick due to heavy rain and the Autocar was having difficulty gaining traction climbing that very same grade so he climbed out of the cab onto the top of the trailer which was loaded with corn. There were feed sacks underneath the tarp on the trailer and he commenced to pulling out feed sacks and dropping them under the drive wheels to gain enough traction to make it up the grade. Kenny would tell of a mountain grade that was so long and steep he could put the Autocar into a lower gear and have his wife, Vonda, steer while he took a nap. Kenny would often haul corn to Florida and bring back seashells. I think he did this for the local grain elevator as they would grind up the sea shells into their chicken feed. On one of these Florida trips, he brought back two live baby alligators which he left in the family bath tub while he continued trucking. As I recall these alligators were about a foot long. My aunt Vonda told me they went to the neighbors to take a bath while the alligators occupied their bath tub. Kenny told me that he purchased a load of watermelons which he and Vonda hauled out to Tucson, Arizona, where Vonda's parents lived. His plan was to sell the watermelons in Tucson to pay for their trip. Apparently, watermelons were not in vogue right then as he had some trouble unloading them. He told me the visit stretched out longer than he had planned and he was concerned he might have to take up permanent residency in Tucson.

Kenny traded the Autocar in for an International Scout in 1961. The dealer had to rebuild the Cummins engine due to a rod knock and then sold it to a local farm. Uncle Joe said you would see it going down the road hauling grain.

Kenny and Joe were also in the dozer business during this time frame. They owned an International TD9 bulldozer. I only have a few memories of their dozer. I remember watching Joe dig a house basement with it. And I remember Kenny dropping it off and Scotland and my grandpa using it to dig the basement for the addition he added to his house. Prior to the addition, my grandparents had an outhouse, no inside plumbing, and no basement, although it did have a coal bin.

An International TD9 bulldozer, although not the exact same one owned by Kenny and Joe.

I think Kenny first and foremost considered himself as an Ironworker. He worked at all the various duties Ironworkers perform when erecting a building; he was a connector, rigger, rodbuster, bolt up gang, decking gang, welder, you name it. He built various structures throughout the Midwest even traveling to Alaska to help construct the trans Alaskan oil pipeline.

One of his more interesting assignments was in the assembly of a 145 cubic-yard dragline for Peabody Coal Company at its Dugger, Indiana, pit in 1966. At the time it was the largest dragline of its type in the world and had to be transported to the site on a convoy of trucks where it would be assembled. Kenny was part of the crew that assembled this massive machine. The son of the superintendent for Peabody Coal company was working at that site during his summer break from college. This man became a Pentecostal Preacher later in life and my Uncle Joe joined his congregation and they became friends. This preacher performed the service at my Uncle Joe's funeral. At the service he brought up his employment for Peabody Coal Company and told the attendees that he met his first "Asdell" while working at the Peabody strip mine.
He said he had "screwed up" and his dad was "chewing him out" when this crazy little red Volkswagen Beetle came roaring down the drive raising dust and suddenly screeched to a halt. Pastor Greg said his dad lost interest in lecturing him and focused on the intruder instead. Once stopped, the driver's door opened on the Beetle and out stepped this giant of a man with a full bushy beard sporting an

unruly head of hair holding an M-16 rifle, which he proceeded to fire into a large pile of gravel. (Pastor Greg identified the rifle as an M-16 but was most likely an AR-15. Although, Kenny had the "ability" to acquire items unavailable to most. Per his brother Bill, he had a Thompson Submachine gun at one time). This was Pastor Greg's first introduction to Kenny Asdell. I have overheard my wife telling her friends that after hearing the preacher share that story, she was convinced that the Kenny Asdell stories she had heard previously were undoubtedly true.

I worked with Kenny on an addition to the Ameritech building in downtown Indianapolis in 1973 and was amazed at how many of the ironworkers made a point to talk to him. He could definitely draw a crowd. ***My personal Ironworker experience is included farther back in this book. ***

Kenny and a fellow Ironworker made a trip to Mackinaw City, Michigan, when construction began on the Mackinac Bridge but the project was just getting underway and there wasn't need for extra manpower at the time so they headed back to Indiana.

Upon returning home he got on a crew building a free-standing television tower close to Weir Cook Airport in Indianapolis that, upon completion, was the tallest free-standing tower in the world. It topped out somewhere in the neighborhood of 1,050 feet. Kenny said planes were coming in for a landing below them as they got close to completion. The top section was set with a helicopter.

Kenny worked on Assembly Hall at Indiana University which turned out to be the only job on which he took a fall. He slipped on some loose concrete and lost his footing. It was well over 100 ft. to the ground from where he fell but, luckily, he latched onto a rope that was suspended overhead which slowed his descent and redirected his route causing him to land on a sheet of plywood left by electricians which held their electrical boxes, conduit, wiring, etc. That rope saved his life. He suffered some nasty rope burns on his hands, some bruises, and an injury to his ankle.

Assembly Hall.

When Kenny began ironworking, they used rivets to connect beams, rather than the bolts they use today. That process required a four-man crew. The rivets were heated up and when they reached the appropriate temperature, the thrower removed it with long tongs and using the tongs tossed it to the catcher who caught it in a tin scoop. The "bucker upper" using shorter tongs would take the rivet and place it in the hole whereby the riveter would set the rivet using a pneumatic rivet setter.

These rivets were red hot (**see picture of rivets being heated**) and this operation was occurring while these guys were working high up in the air without any type of harness requiring the thrower and the catcher to be very accurate. Late in his life, Kenny told me that he always wanted to set up a riveting crew and perform demonstrations at fairs and festivals. He said he had enjoyed the sight of hot rivets flying through the air when it was getting toward dusk as they would leave a trail of sparks in their wake. They actually do have riveting demonstration teams. One place where you can see them is at the annual Ironworkers Festival in Mackinaw City, Michigan.

Example of beams connected by rivets. Imagine how much time and labor was required to install all of these rivets. Plus the danger of getting hit by a red hot rivet or falling while attempting to catch a poorly thrown one. If you ever visit the Mackinac Bridge, observe the rivets holding it together, there are millions of them.

55 Year Members
BM Jeff Stinson, Kenneth Asdell, Harry Fryer, and BA Earnest Thompson.

Kenny receiving pin for 55-year membership in Ironworkers Local 22. He received this award in 2008.

Kenny's son Randy, third from left, receiving pin for 30-year membership in Ironworker Local 22. He received this award in 2008.

During the first Gulf War, after Saddam's forces had set all the oil wells on fire in Kuwait, Kenny visited the Army Recruiter in Bloomington and offered his services in bringing these oil wells back on line. He told them he had years of experience in oil fields, oil drilling, and rigging and would work for free if they could get him over there. Kenny was about 64 years old at that time. Kenny did drill some oil wells in his life, plus his brother Joe worked on some oil wells as well. These wells were all in southern Indiana.

Kenny traveled to Alaska and got on the trans-Alaska pipeline project. He worked in Prudhoe Bay and stayed at a company provided camp, called "Crazy Horse Camp". He welded on supporting structures for the pipeline including buildings. He said they worked twelve-hour shifts, 7 days a week. Nothing ever shut down, trucks and equipment were never shut down due to the extreme low temperatures. He wore down filled coveralls and mittens which were provided by the contractor although the recipients all had to pay for them. His family displayed Kenny's down filled coveralls at his memorial service.

Crazy Horse Oil Pipeline Construction Camp Prudhoe Bay 1970

Kenny brought home an art piece that was fabricated from a section of the actual pipeline cut into the shape of Alaska. His was inlaid with stainless steel and blued with a torch. The route of the pipeline was "prick punched" on the surface. I don't know what he ever did with it. Apparently, that was a common form of artwork that the workers were creating at the time. And they were selling at a premium.

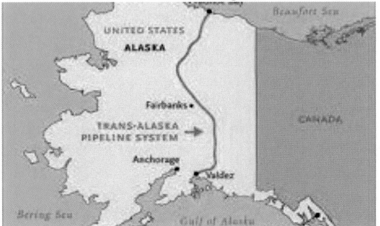

Route of Trans-Alaska Pipeline; Kenny worked at the very top of the map, Prudhoe Bay.

Kenny said when he was working on the pipeline there was another guy working with them that was bigger than Kenny, so their coworkers called that guy "Big Heavy" and they called Kenny "Little Heavy".

Big Heavy had been a ship builder (welder) at a ship yard on the east coast but had quit and traveled to Alaska to pan for gold. When the pipeline started hiring, "Big Heavy" signed on to build up his "grubstake".

As if being a truck driver and ironworker were not enough, Kenny was also a member of the Operators Engineers Union. The ironworker union shared a business agent with the operating engineers union and Kenny was friends with the guy so he signed up with the operating engineers. To my knowledge, he operated pan scrapers exclusively (sometimes called turnapulls). Joe told me that when Kenny reported for his first job, he had told the job foreman that he had extensive experience operating a scraper. The foreman led him to his machine and Kenny climbed into the cab and called out to the guy; "hey buddy, how do you start this thing". The foreman said "I thought you knew how to operate one"? Kenny replied, "yeah but it has been a long time". That takes "cojones". Kenny must have been a "quick study".

Label it a scraper, pan scraper, or turn a pull, take your choice. This is the make & model of the machine Kenny operated. Most of the time he was scraping off top soil in preparation for strip mining (for coal). The top soil was saved and reused in the reclamation process afterward.

Kenny owned two different taverns (or "clubs" as he referred to them) one in Crane and one in Switz City, Indiana. The Crane tavern burned down. He later purchased the business in Switz City. The Switz City joint they name the "Oodle Inn". I think Kenny's specialization was in "crowd control" as he was involved in a few brawls during his ownership. Kenny's son, Randy, and I accompanied him to the tavern one day to help clean up the place which would put a little money in our pockets.

The barmaid told us to ask Kenny about the fight they had the previous night. When we asked Kenny what happened, he shrugged and said, "not much". Then he added; "I went bam bam (motioning with his fists) and it was all over". We asked who won and his reply was "you don't see no dirt on my back". End of discussion.

Memories & Stories from Which Legends Are Created

Many years ago, Kenny was the owner an Indian motorcycle. Joe shared with me that he had ridden Kenny's motorcycle and noticed it had a significant vibration at high speed, mostly when deaccelerating. He told Kenny that he should have the vibration checked out before there was a tragedy. Kenny responded that "Joe just didn't know how to ride it". Joe said it wasn't long and some guys brought Kenny home in the bed of a pickup. The motorcycle had started vibrating violently and Kenny had to lay it down at a high speed on the highway. Kenny had slid down the pavement and sustained significant road rash. His clothes had worn through, even his wallet had a hole worn in it, but no broken bones.

Kenny's Indian motorcycle with younger brother Bill onboard. At Kenny's memorial service, Bill shared a memory about riding with Kenny on the Indian motorcycle. He said local motorcycle guys had a place they would meet and "shoot the bull". Kenny had invited Bill to join him one night so Bill climbed on back of the Indian and off they went.

Bill said it was terrifying, Kenny never backed off the throttle and when cornering he leaned the bike so far Bill said I swear I could see sparks where the foot pegs scraped the pavement. Over his lifetime, Kenny owned several motorcycles, as he loved to "wheel and deal". I remember a BMW he had picked up. It had the famous BMW "opposed cylinder" configuration and was shaft drive. In those days, BMW was considered the "smoothest" motorcycle one could ride. The "king of touring bikes".

The last motorcycle I remember him owning was a Kawasaki, not sure of the model, maybe a 650 or 750. I rode it and didn't like the handle bars, seemed they were set at an odd angle and was hard to handle at low speeds. Kenny didn't seem bothered by them.

I was told by Brent Dillman of the Snapper Shop in Scotland that Kenny laid it down going around the corner in front of his shop. At that time Kenny must have been in his early 70's.

When we were kids back in the 1960's, Kenny purchased some Honda Trail 90's. The Trail 90 was a step through model with dual sprockets, one large one for the trails and a small one for the road. They also utilized a centrifugal type clutch so had no hand clutch on them. If you set the bike up with the large trail sprocket you could climb a mountain, and Kenny did just that down in Rocky Branch. He busted the frame on one of the little Honda's by jumping it up over a log. They were a tough machine but they may have met their match when they encountered Kenny.

In more recent times, Kenny kept a "stable" of three-wheelers and then four-wheelers. My favorite was a Polaris diesel four-wheeler he owned. Those diesel four-wheelers were as rare as "hens' teeth" which was not surprising as they were not in high demand since they were noisy, smelly, and slow. If you had hopes of seeing deer while on a four-wheel journey, you best not take the diesel. But that thing did have torque, I swear it could climb a mountain. Trips through Rocky Branch demanded a four-wheeler with good traction (four-wheel drive) and torque.

I made a trip to southern Indiana with an old friend from Michigan Department of Transportation, Dave Budd, and my stepdaughter, Mariah, to enjoy some trail riding on four-wheelers. Dave brought a Polaris side by side and a four-wheeler. The first afternoon, we made a journey through Rocky Branch back to First Creek, which was swollen so we didn't attempt a crossing. The next day Kenny joined us as our tour guide. We had figured he might ride the side by side as he was in his 80's at that time. But of course, he rode his own machine, a three-wheeler which he told us was easier for him to mount than a four-wheeler.

He took us on a spin though Rocky Branch, then north of Scotland to the old Blackmore homeplace where he was born and to the Hasler family cemetery which is where some of our ancestors are buried.

The patriarch being Peter Hasler who was the great-grandfather of my Grandfather Drib Asdell on his mother's side (Clara Stone). Peter Hasler migrated from Switzerland. The trip involved several stops while Kenny shared historical information and stories in general. It was an educational and interesting trip.

Dave Budd, Mariah, and Kenny on the day of our four-wheeler excursion

Mariah and Dave Budd descending the hill down into Rocky Branch

Mariah with Kenny's dog "Skunk" riding on the Polaris diesel machine. Whenever the four-wheeler left the yard "Skunk" would be on it. He rode with Kenny when we made the journey north of Scotland. We stopped at the gas station and the kids were ecstatic when they noticed 'Skunk" riding behind Kenny.

Rocky Branch

Mariah and my son Seth, under rock outcropping while riding in Rocky Branch many years ago.

Kenny's memorial in Rocky Branch carved by his grandson Josh Asdell.

This picture was taken on the day of our journey right after Kenny had shown us where he shot himself through the leg. Seems he found a little pistol from his tavern days and when he tried to work the action found out it was frozen. During his attempt to free it up it went off, apparently it had been setting around loaded. The bullet went though his calf doing little damage but he went the to hospital for treatment. The hospital has to report gunshot wounds to the police. When the state police arrived, Kenny told them "don't tell Obama, he'll take my gun away". As he was telling the story he was waving the pistol around so I told him to put the pistol away before he shot someone. He told me it wasn't loaded. I said, "do you know how many people have been shot by unloaded guns"? Or at least they thought they were unloaded. Dave Budd later told me that after hearing the gun story from Kenny he decided that the Kenny stories I had shared with him over the years were indeed true.

The ashes of Kenny, my mother June, and stepfather Jack Lester, were spread at Rocky Branch. Rocky Branch has been part of the Asdell family since my greatgrandfather, Dave Asdell, purchased the farm on which it is located in 1898 or thereabout.

Dave & Clara Asdell at their farm. 1898 or thereabout. The farm where Kenny resided until his death in 2016.

Bill on Kenny's horse, named Red. Kenny and Joe were both horsemen. Kenny always had several horses and ponies for the kids. We would have trail rides through Rocky Branch. Traveling down the branch was "tricky" due to the slick rock bottom and loose stone (this also applies to four-wheeler rides). Every kid that ever rode a horse or pony at Kenny's was taught the lesson by Kenny; "if you get bucked off, you get right back on". I remember when I was a little boy, Kenny coming into the house wearing a white tee shirt that was covered in dirt and hoof prints. Seeing that, my eyes probably got as big as "silver dollars". When asked what happened Kenny replied as a matter of fact; "his horse threw him and stomped him". That incident didn't stop Kenny from getting back in the saddle.

Kenny and Joe on horseback, early 1960's. They had just raced each other in the field across the road. Jack Lester, my stepdad, used to talk about watching Kenny pick up a Shetland pony and load it in the back of his pickup. Kenny wanted to take this pony to Scotland for the kids to ride and needed to load it in his truck. When Jack asked him how he was going to load it, he said I'll just pick it up and put it in there. Jack said that is exactly what he did after he caught it by the tail and the pony pulled him off his feet and drug him around the barn lot until becoming exhausted. Kenny got up off the ground, picked up the tired pony, and stuffed it in his truck.

Kenny and Joe used to catch horses by the tail all the time. Horses at Kenny's farm were almost free-range horses due to the amount of acreage they could roam so catching them could be challenging at times.

Joe on horseback.

Kenny and daughter Sheryl on horseback

Joe Asdell

Kenny and Joe, as well as the rest of their siblings, spent a good bit of time at their grandparents farm, Dave & Clara Asdell, during their childhood. This was the farm that Kenny later purchased. Upon Dave Asdell's death in 1953, the farm was divided up amongst his heirs. Kenny purchased the pieces of the original farm from all the heirs and restored it to it's orginal form. Dave Asdell farmed with horses. Joe told me that his grandpa purchased a team from "out west" that was barely broke to drive and not broke to ride at all. His grandpa told Kenny and him to stay away from those horses for fear they might be injured. Joe said that he and Kenny would climb up into the hayloft and get above these horses when they were in their stalls. As the horses ate their hay and grain, he and Kenny would lower themselves on the horses backs. In the beginning the horses were nervous and "fidgety" but over time didn't pay any attention to the boys on their backs. Joe said their grandpa had the team in harness and headed to the field when Kenny and himself decided it was time for them to make their "debut". Over the protestations of their grandfather they jumped up on the back of the horses. The horses were now accustomed to having the boys on their backs so they did not even flinch. Joe said his grandfather was dumbfounded. He said after that event they rode those horses whenever they felt the urge. My Uncle Ben told me about a mare his grandpa had named "Nell". He told me old Nell was blind. He also told about riding Nell while Kenny and Joe rode two other work horses, maybe the two mentioned above? He said Nell was slow and didn't keep up with the other two horses so Kenny and Joe made him switch mounts giving him one of the more "spirited" animals. Must have been challenging being the younger brother?

My grandpa Drib Asdell was always bewildered as to Kenny and Joe's attraction to horses. He said when he was young you only had two modes of travel; walk or ride a horse. He didn't care for horses so he would normally walk. He figured horses were a necessary "evil" on the farm, as they were needed for field work, but he didn't get any enjoyment out of riding one. I have that very same sentiment about snowmobiles. We moved to Michigan when the snowmobile craze was in full force.

I rode snowmobiles back then occasionally but found those early snowmobiles were not all that dependable and breakdowns were common. Plus I never had money to purchase a quality snowmobile suit so generally froze my "butt off". Nothing more fun than being stranded on a frigid winter's night in the middle of nowhere while you suffer frostbite. I only mention this because sometime in the mid 70's I took some friends to southern Indiana in early May. We were heading toward my Uncle Kenny's farm and off in the distance we spied a snowmobile traveling down the road. Seeing a flying saucer would not have been more surprising. No one, and I mean literally no one in southern Indiana had a snowmobile, I did not even know where you could buy one down there. Plus, no one, in Michigan or Indiana, rides a snowmobile down a paved road in May. My friends declared something to the affect of "look at that crazy guy on the snowmobile". I stated matter-of-factly, that crazy guy is my uncle. He was too far away for me to recognize and my friends pointed that out. I told them, I don't have to recognize him, there is only one person that would (a.) purchase a snowmobile down here where you receive hardly no snow and (b.) run it down a paved road in May. Sure enough, we caught up with the snowmobile which was being operated by Kenny.

The Menagerie

I have already mentioned the two alligators Kenny brought home from Florida and housed in the bathtub, much to the dismay of my aunt Vonda, but those two alligators are only the tip of the ice berg. Kenny and his family had a plethora of "pets" through the years, some which are common and some quite unique. There were always dogs at Kenny's place and they always roamed free, he never believed in confining a dog. His dogs were of all types and descriptions but generally larger breeds. He had several German Shepards over the years. The first and most memorable being Duke who was absolutely surly. I believe that dog bit everyone on the place except for Kenny. My cousins raised a runt pig which the neighbor gave them. It was allowed to run loose with the dogs. It imprinted with the dogs apparently. It was getting rather large so Kenny sent it back to the neighbor who promised to use it for breeding purposes only and not to send it to slaughter, much to the relief of Kenny's kids; Randy, Dee, and Sheryl.

There was only one hiccup, that pig had grown up with dogs his entire life and "apparently" believed itself to be a dog. It would have nothing to do with the other pigs which, sadly, was an unfortunate situation for the pig as he was not suitable for breeding. Fred and Ethel Ziffle had nothing over Kenny and his family.

Kenny and my cousins had pet raccoons and pet squirrels which was not so uncommon. He also had several whitetail deer as pets over the years. Several that stayed in the house. My cousin Sheryl has pointed out the irony of her Dad's deer being allowed in the house when dogs were banned and had to remain outdoors.

In the picture is my son Boone in Kenny's kitchen sharing peanut butter with Kenny's pet deer "Buck". I can remember walking into the den where Kenny would be napping and Buck would be lying on the floor next to the couch enjoying the cool breeze from a fan.

Kenny would take his deer for rides in the car stopping in at businesses. I remember his taking Buck to the John Deere Ag dealer where he scored a free hat. I believe he took one to a tavern at least once. He didn't lead the deer around, they just followed him like a puppy. When he had the red Volkswagon Beetle, he would load up his deer and head to town. That was a sight to behold.

The crème de la crème of pets had to be Satan. Satan was a mountain lion (or cougar, puma, whichever you prefer). He acquired Satan from a woman who raised him from a cub.

Satan, the mountain lion. Kneeling in the picture is Kenny's cousin, Ben Osborn. When Satan gave you that "cat stare" it would send shivers down your spine although I never witnessed him being especially aggressive, not that I wanted to.

Kenny kept Satan in front of the house by the pond. I remember my Grandpa Drib talking about fishing on the pond and had caught a stringer full of fish. It was a shorter walk to his car around Satan's side of the pond so he checked to see where Satan was at and didn't see him so that is the route he took. My Grandpa said when he got to within Satans domain he glanced down and the cat was following in his footsteps eyeing the fish. Papaw Drib said he never heard that cat and wouldn't have know he was next to him if he hadn't of glanced down. Papaw Drib said "I told Satan if you want those fish they are all yours". He didn't actually give them to the cat and the cat never tried to take them. He was just curious, I guess.

I was teasing Satan with a hat and he was swiping at it with his paw. His claws were retracted all the time we were playing around so he wasn't being aggressive, he was just playing. But, he did catch my hand with the tip of a claw and left a jagged scratch the length of my hand. It was a shallow scratch but nasty looking. No big deal, Satan was not trying to slash my hand off, if he had been, I would be missing a hand.

The funny part of the story was when I was sitting on a bar stool at a tavern in Cedar Springs, Michigan, and the guy next to me asked what I had done to my hand. I knew right then this was going to be good. I casually told him, I got scratched by a mountain lion. I am sure you can imagine the guffaws that guy emitted. I left it at that thinking,"the one time I am telling the truth but no one will ever believe me". In their defense, they didn't know Kenny.

Satan, the cougar, riding on the deck of the Navy DUKW (more commonly known as "Duck") that Kenny owned. Cousin Sheryl is sitting on the other side of Satan, guess she was "cat sitting" The Navy DUKW is a six wheel drive amphibious modification of a 2-1/2 ton truck used in WW II and the Korean War. There was a company in Branson, Missouri, that had a fleet of them and offered rides until one sunk. This picture was taken in Scotland, Indiana, during the Scotland Festival Parade.

Kenny owned this "Duck" and also owned an Army "Weasel". The Weasel was a tracked armored vehicle originally designed for use in snow, specifically for an operation in Norway.

The Norway operation was scrapped but the Army found it perfectly suited for other uses. Army Weasels came ashore with the troops on "D" Day. Army Weasels were manufactured by Studebaker in South Bend, Indiana.

My Uncle Joe had a pet raccoon named Romeo which he acquired from my Uncle Bill. Romeo was never confined and always ran loose. He was allowed in the house which I found surprising as my Aunt Mary kept a very tidy house and never did I think she would allow a raccoon inside her home. But, Mary liked Romeo as well. As raccoons mature they tend to go exploring and will be gone for increasingly longer periods of time until they no longer return. Romeo, true to form, followed that same pattern. One day Mary was returning from town during the daytime and noticed a large raccoon run across the road and begin climbing a telephone pole. Mary said she figured it was Romeo so she stopped the car and walked over to the telephone pole and said 'Romeo, you get yourself down here". Romeo came down the pole, followed Mary to the car, and climbed inside when she opened the door. Mary took Romeo back to her house and fed him. The next day he took off never to return.

Joe's daughter, Jo Ellen, had a Shetland pony that was colored like a Pinto horse. The little pony's name was Rusty. When they purchased Rusty he was not broke. Uncle Joe and Jo Ellen broke Rusty to the saddle and they taught him several tricks. Jo Ellen taught him to count. Uncle Joe taught him to kneel down like a camel. I don't remember if you could mount Rusty while he was kneeling as I don't know if he was strong enough to get up with a rider on his back.

Joe was a houndsman. He had foxhounds up until he got burned and sometime after that switched over to coonhounds. Two of his early legendary hounds were Rowdy, a Treeing Walker, and Blue, a Bluetick Walker foxhound cross. When Rowdy got old, Joe left him in Scotland so Bill and Papaw Drib could hunt him. Rowdy would follow Bill wherever he went so if Bill didn't want Rowdy following him he would sneak out of the house. Whenever Rowdy would discover that Bill was missing he would follow his trail. Townsfolk would tell about seeing Old Rowdy standing next to the pavement where he had lost Bill's track, bawling mournfully.

Joe had a Treeing Walker hound named Doc who took up residence in Joe's pickup. Doc would ride with Joe to town when Joe was working at the Blacksmith Shop and hang around the truck all day. If Doc got hungry he would raid a trash can. Joe said many times he would get ready to leave work and there would be a trash can next to his truck that Doc had drug back. I guess it was his "lunch box". Once when Joe's mother-in-law, Pearl Crane, used his truck to pickup chicken feed, Doc saw the truck heading out of the driveway and came at a dead run jumping through the window into the cab almost landing on Pearl's lap. When Joe replaced the truck he retired it to the woods behind his barn. He said Doc stayed in the old truck for a few days before he finally changed residences and moved into the new truck.

Mark standing next to 1951 Ford F-3 pickup belonging to his son Boone. Uncle Joe's red 1952 Ford F-1 pickup looked identical to the truck in the picture. This was the pickup in which old Doc, took up residence.

Kenny and his wife Vonda (Beasley)

Joe and his wife Mary Ellen (Crane)

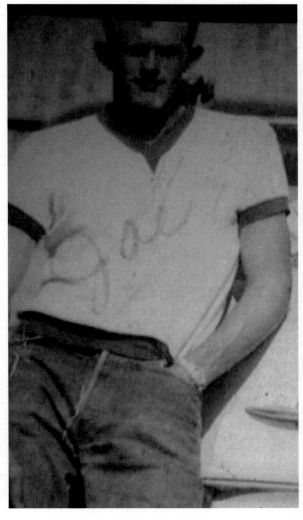

Young Joe Asdell. I would surmise this picture was taken right around the time Joe was building silos in Michigan. After recuperating from his burns, he married Mary Ellen Crane in 1952. Mary's parents were dairy farmers just north of Raglesville, Indiana. Joe partnered with his father-in-law on the dairy farm plus share cropped additional acreage for cash crops. Joe's father-in-law, Hasel Crane, milked 30 or so head of mostly Hosteins. He ran a Grade A dairy operation and Joe and himself were considering expanding their dairy operation when disaster struck and their barn burned to the ground in 1967. It was almost a "double disaster" as Joe had just purchased an 806 Farmall tractor, the first diesel and largest tractor he had owned up to that time and the tractor was parked under the barn lean-to. Luckily, they got it out in time. After the barn fire, Joe and Hasel made the decision to discontinue milking cows.

Farmall 806 tractor. Identical to the tractor that was almost lost in the barn fire.

Joe worked for wages while he was farming. He worked as a union cement laborer at one spell. He worked on a project at Mitchell, Indiana, which, upon completion, was to be a manufucturer of precast concrete structural beams. Kenny worked on the same job as an Ironworker. Joe said it was very cold and they were working out in the open terrain with nothing to temper the cold winds. He said Kenny was welding high in the air out in the open with no protection from the bitterly cold wind. Joe said I actually felt sorry for him, but in hindsight, I should have known better. It didn't take Kenny long and he was promoted to foreman spending most of his time in the job shack next to the stove. After that, Joe said, "I was the one out in the bitterly cold wind. I doubt Kenny felt much sorrow for me."

Frozen ground and heavy cement mixer traffic caused a water main to burst in the town of Mitchell. The contractor that Joe was working for asked for volunteers to stay and repair the water main. Joe offered to stay and help. It was winter, of course, and the crew had to work in freezing water. Joe said it took them all night to get the water main repaired and that was after they had worked a full shift the day before. They worked 24 hours straight through after which they were frozen and exhausted. Joe offered me some tidbits on how to keep going without sleep which served me well at Michigan Department of Transporation when I worked a 24 shift during a winter blizzard after which I was also frozen and exhausted. The only tidbit Joe did not share was how to stay awake on the drive home.

Joe told me a story about a job he was on where they were working several stories in the air. They used a crane to hoist the cement up in what Joe referred to as a "georgia buggy" whereby the laborers would guide it into place and trip the door to empty it. Joe said the crane operator raised the unit prematurely while Joe was still emptying it, trapping his hand. He said, "I thought I was going to lose my hand so I grabbed ahold with my other hand and rode it to the ground". When he reached the ground the crane operator and the job foreman were livid, they thought he was "playing games". He said no matter, "I went home with both hands".

This is the type of unit Joe referred to as a "Georgia Buggy" which trapped his hand and he rode off the building to the ground. I have always heard it called a "concrete hopper".

First tractor Joe owned. Farmall Super MTA. He traded it on the 806.

Pugilism

I have mentioned a few of the physical altercations that Kenny and Joe participated in but there are many more even to the point where they have become a subject of legend.

- My Uncle Bill shared a story that involves his late brother-in-law, Clifton Miller. Apparently, Clifton stopped at a tavern in Bloomfield, Indiana, located on the square for a quick beer. While he was inside a stranger walked in who was loud, boisterous, and downright obnoxious. Clifton thought to himself, "I would like to see Kenny Asdell walk in about now". Clifton said as he left the bar Kenny was coming in. He figured this is about to get interesting so he sat down on a bench outside the tavern and waited. He said he didn't have to wait long and here came the obnoxious stranger out the plate glass window onto the sidewalk.
- Joe told me that Kenny was tending bar in Crane around Kentucky Derby time. It was a slow night and there was an old World War I vet sitting at the bar visiting with Kenny when in walked an out of town truck driver who had a load to deliver to the Depot. I do not remember the exact details but the truck driver got "out of hand" and Kenny tossed him out. The truck driver

went to his rig and came back with a knife with which he attacked Kenny. Kenny managed to disarm the guy and put him down. The truck driver began crawling for the door. As he was crawling, Kenny would kick him in the rear end with his cowboy boots so hard it raised the guy off the floor. Each time after he got kicked the truck driver would crawl faster. The old guy at the bar was guffawing and hollered out to Kenny; "put a saddle on that guy and we will enter him in the derby, that's the fastest thing I have seen in a long time".

- My Aunt Mary told us about a time when Joe came home after he was attacked with a knife wielding opponent. I don't remember all the details of that story but I do remember Joe responding, "well the guy never got me". And Mary retorting "maybe, but not for the want of trying, he cut your shirt to ribbons".
- Uncle Bill told about a competition coonhunt in which he was participating in northern Illinois. He said one of the hunters in the cast was a belligerent argumentative cuss and making the other particpants frustrated. Bill said at the end of the hunt when they all had to sign the scorecard, he signed and then handed the card to this guy. The guy looked at the card and asked "your name Asdell"? Bill said "yeah, why"? The guy asked him if he was any relation to Kenny Asdell and Bill replied that he indeed was, Kenny was his big brother. The guy said "I have ironworked with him. He is the meanest man in Indiana". Bill said whenever he ran into that guy after that revelation, the guy caused him no more grief.
- Kenny would visit Joe's in-laws, Hasel and Pearl Crane, at their farm and join them for dinner if he wasn't working. Pearl was a great cook, as well as her daughter my Aunt Mary, and always did a complete course at every meal. So Kenny wasn't about to pass that up. Apparently, Hasel found out the neighbor guy had shot and killed his farm dog. Hasel told Pearl, "the next time Kenny comes for dinner I am going to have him go down there and kick that guy's ass".

- Uncle Ben shared a story about Kenny and a couple guys that stopped in at an intersection. They stood in frotn of his car, one guy on each side. Ben was in the passenger seat and he said "I just reached over and pushed down the door lock. I didn't want any part of this. Kenny jumped out of the car and confronted the guy on the driver's side. Kenny took him out with one punch and then ran around to the passenger side and dispatched that guy as well. Ben said when Kenny got back in the car he turned to him ans asked, "what is wrong with people anyway"?
- Another Uncle Bill story; he was dropping his future wife, Sandy, off at her home in Odon, Indiana, after a date when he was approached by a friend who told him he needed to get down to the Odon Tavern right away. His brother Kenny was inside and had been involved in a fight and the town marshall was waiting across the street for Kenny to come out. Bill said he hustled over there and went in the back door. He said there were three guys laying on the floor, blood everywhere, Kenny was leaning against the bar, intoxicated, bruised, and bleeding. He said Kenny was sobbing and blubbering "I thought he was my friend", referencing the bartender. Apparently two guys jumped Kenny and he turned his back to the bar to fight them off. While he had his back turned the bar tender thumped in the back of the head with a blackjack a couple of time. Kenny drug the bartender from behind the bar and dispatched him. Bill said I got him out of there and took him home.
- Uncle Joe (and Kenny himself) told me a story that occurred soon after Kenny returned from Alaska. Kenny had an acquantance who was an attorney and tavern owner (referred to by Kenny as a club). This guy was having problems with a motorcycle gang who were driving off his clientele so he approached Kenny for his help. Kenny went to this guys joint as a bouncer. Sooner than later, he tangled with some of the biker gang members. One of them struck him across the cheek with a handful of pool cue sticks fracturing Kenny's jaw but not putting Kenny down. Kenny wrested the cue sticks from the guy and proceeded to beat them with a cue stick. The police showed up and arrested the guys. Kenny told the club owner that he guessed he was done with that job. Kenny was somewhere in his mid-40's when this occurred.

After my Uncle Joe had passed away, Kenny and I were both at Joe and Mary's home visiting out on the porch when Kenny told me, matter-of-factly; "you know Mark, I have probably whipped a thousand men in my life, made a better person out of everyone of them". That just made me chuckle so I responded; "well you know Kenny, I will bet none of them would agree with that statement". You gotta love him.

Kenny sitting on his own porch

In 2004, my Uncle Joe turned 75, my Aunt Mary Ellen, and my Mom turned 70. Joe's birthday was in August, Mary's was in September, and Mom's was in November. I was in down there in September so I bought them a birthday cake at Walmart and had it decorated with their names and ages. I invited Kenny down for ice cream and cake. It was a really nice time. Mary told me she could have baked the cake and saved the money. Which I said was silly since it was for her birthday and then she told me she have never had a birthday cake before. That blew my mind but for sure by the time you turn 70 you are entitled to a birthday cake.

In 2006, we held a "Drib & Sophie Asdell Descendants" reunion at West Boggs Park, in part for my Uncle Joe as he was fighting cancer and was feeling pretty good so thought it would be really great if he could see everyone. We had a really could turn out. In less than two years Joe was no longer with us.

Joe at 2006 reunion

2006 "Decendants of Drib & Sophie Asdell" Reunion.

West Boggs Park

Loogootee, IN

Joe, Kenny, Mom (June) and Bill at 2006 reunion

When Joe was racked with cancer and couldn't even stand up unassisted, let alone walk, he told me: "I could still farm if only someone would help me up on the tractor seat". No quit there.

Kenny was driving semi truck across country when he was in his 70's. He was making runs regularly to the west coast. The last truck driving gig he got was for a local farmer hauling grain. I don't know how old he was but had to be pushing 80. He had so much difficulty getting into the truck cab that the farmer never called him back to drive. Give him credit, he did get in one day.

After Joe lost the ability to walk unassisted, he would have Mary Ellen park his zero turn mower by his back porch so he could access it. He would use his walker to reach his mower and then pull himself onto the seat. Three times, that I am aware of, he hung up the mower next to his pond. One time when it happened I was there and pulled him out with his four wheeler. The other two times he started crawling to the house. One of those times a passerby saw him and picked him up in his pickup. The other time a neighbor saw him and helped him to the house. He never quit.

Mary & Joe Asdell

Kenny

World War II Honor Flight

Kenny drawing a beer for my sisters; Kathryn & Ruth at one of his infamous 4th of July birthday parties

Joe, Ben, Mom (June), Bill, & Kenny @ one of Kenny's infamous 4th of July birthday parties   (1986?)

Mary Ellen, Joe, Cindy, & Dave at the 4th of July birthday party

Joe, Ben, Mom (June), Bill (in back), & Kenny

Ben's 70th birthday party

Joe and Jo Ellen

Mary Ellen & Vonda east of Scotland searching for a Christmas Tree. Dee Dee & Jo Ellen are with them.

Kenny, Mom (June), Bill, & Ben

Scotland School Reunion

"Some people die at 25 but are not buried until 75". This quote is attributed to Ben Franklin. This certainly doesn't apply to Kenny and Joe.

"Either write something worth reading or do something worth writing". Another Ben Franklin quote. I believe they certainly lived a life worth writing, hence why I wrote about it. I hope what I wrote is worth reading.

Life is a continuous circle.

My great-aunt, Peg Asdell, lived on the home place with her father, my great-grandfather Dave Asdell, until he passed away.

Peg Asdell and her Dad, Dave Asdell

1952 (the year before Dave passed away)

Many years later, Kenny, grandson of Dave Asdell, lived in the very same home with his daughter, Dee, until he passed away.

Kenny

Dee

When Joe lay in a hospital bed suffering from serious burns, Kenny was there. As much as he could be present, he was with his brother who was literally fighting for his life.

When Joe was confined to a chair, cancer consuming his body, Kenny would be there. He would sit with his brother Joe. They would share stories and memories allowing Mary Ellen, Joe's primary caregiver, the freedom to leave the house.

After the premature death of my father, when my mother was struggling Kenny was there. He would stop in and visit with his sister and drop off groceries for us. When the insurance agent tried to prevent my mom from collecting on my Dad's life insurance, Kenny was there. He and Ben paid a visit to the insurance agent in my mom's behalf, her sister, and whatever they did, worked as my mom received her settlement. When I, as a little boy, lost my Dad, Kenny was there. He showed up regularly and took me for a ride in the old Autocar, sharing his time with me.

Heroes

I was honored a few weeks ago by being chosen to deliver the eulogy at the memorial service for my mom's oldest brother, Kenny Asdell. When I was asked by my cousin if I could or would do this, there was no way I could decline. When I began the service I shared the story, it went as follows:

Almost 60 years ago, in a small town not far removed from here; there lived a little boy who had just lost his Dad. His Dad had succumbed to the insidious disease of cancer at the very young age of 28. This little boy shared this tragic loss with his mother, who at 23 was far too young to be a widow, and his two younger sisters who were too young to ever remember their Dad. This occurred back in the time when there were no forms of government assistance, no welfare, no food stmaps……….it did not exist. This small family was left mostly to their own devices, although there was some help offered by family and friends. The little boy's dad had been self-employed in the refrigeration & air conditioning industry and some of his customers (grocery stores and meat cutters) helped the grieving family. Additionally, there was help from the local American Legion Post.

This was all greatly appreciated but there was a need that wasn't addressed. To a little boy, Dad is "bigger than life", a hero who can do no wrong. Dad can run faster, jump higher, out wrestle, out smart, and a constant companion. Tragically, in this instance, Dad was no longer in this little boy's life and would never be again. It was a bleak time. However, as so often occurs, a "super hero" did rise to the occasion. He did not wear a cape nor drive a Batmobile" but rather wore cowboy boots and drove a 1948 Auto Car semi-truck. In case you haven't figured it out yet, the little boy was me. My Uncle Kenny was an independent truck driver back in 1958 and he would stop once a week and deliver a bag of groceries for our family and then take this little boy for a ride in his Auto Car semi-tractor.

For a brief period of time, this little boy could escape from his situation and feel important again as he rode in the cab of a monster truck with his Uncle. To this day I vividly remember my Uncle's cowboy boots working the pedals as he guided the gear shift through all 15 gears while he manned the massive steering wheel found in old trucks not equipped with power steering. For a short time, everything was "right with the world" it seemed. I have worked on and driven many large trucks since that time, but none have been any larger or memorable than that 1948 Auto Car which provided a much-needed respite to one very sad and lonely little boy. I indeed owe a very large debt to my Uncle, so refusing to deliver his eulogy was never an option, but rather an honor.

It is my belief that heroes come in all sizes and all flavors and many influences your life in a positive way without you even noticing. As I have shared with you, I have been blessed throughout my life with many instances where someone has interacted with me and provided knowledge, guidance, comfort or whatever the situation required at the time.

I guess I will leave to the reader whether these occurrences were divinely inspired, providence, coincidence, karma, or just plain old luck. Sometimes it is the smallest of things that make the biggest difference, so never sell yourself short when it comes to helping another.

Joe, Kenny, June (Mom), and Bill Asdell

Ironworker for a day
Mark A Lester

How the hell did I get myself into this predicament, was the thought that flashed through my panic-stricken brain. Better yet, how the hell was I going to get myself out of it? Here I was nine stories up in the stratosphere, standing very much alone, on a steel girder half the width of a sidewalk. I was smack dab in the middle of a twenty-five-foot span between vertical steel columns, which by the way, represented the closest thing to safety that was readily available. My knees had turned to rubber; you could call it the "Gumby effect". My stomach was filled with "butterflies" and my forehead was covered with droplets of sweat. I was real close to a crisis situation. However, I was still coherent enough to realize that there would be no way for a "rubber man" to successfully journey across the beam. I had no choice. So, like they say in the movies, I had to "cowboy up". Call it what you like, but I did successfully make it to the vertical steel column directly in front of me, where I dug my fingernails in for dear life. I also took a minute to tell God how much I really enjoyed it up here in the heavens and was not in any particular hurry to make an express trip back to ground level.

Now that I was safely connected (come to think of it, is this why some Ironworkers are referred to as "connectors"?) to the vertical steel column with a Herculean grip, I paused to reflect on the chain of events leading up to this crisis. It was September 7, 1973; my 20th birthday and I was standing nine stories above downtown Indianapolis, on what was to become a twenty-two story addition to the Indiana Bell Telephone building (now known as Ameritech). This was quickly turning out to be a birthday that I would soon not forget.

This whole chain of events began with an offer extended to me by my uncle, Kenny Asdell, a longtime member of Ironworkers Local 22 in Indianapolis, to introduce me into the field of ironwork (steel erection). Heck, it looked good on paper. Good pay. See the world. Be a man. Wear a hard hat. So, I naively accepted his offer.

I received a telephone call from my uncle on the eve of September 6th and he told me to be at the jobsite the next morning at 7:00 a.m. This was no small feat considering at the time I lived in the vicinity of Grand Rapids, Michigan, and the time of the call was 9:00 p.m. Obviously, I spent a large portion of the night driving.

I arrived at the work site promptly and made connections with my uncle. We first visited the job superintendent's shack where I got signed up and was issued a hard hat. Hard hat, wow, I was one of the guys. Well, okay, I wasn't really an

Ironworker yet but I was starting to look like one, and up to this point, feel like one. Boy was I in for a surprise.

I was told I would be on the "decking crew", which was a good crew for a novice to "cut their teeth" on. Hindsight being what it is, and knowing what I know now, I really believe the "footing crew" would be the best place for a novice to "cut their teeth". But, oh well, water over the bridge.

We all began our ascent up the ladder to the respective floors where we would be working. A ladder that made a perfect right angle to the ground, I might add, one that couldn't be climbed without holding on. Which brings me to **Lesson #1:** Do not hold onto the rungs of the ladder when climbing. I assume readers will know the justification but if not, visit a job site and observe workers departing the building for lunch or at the end of the day. The answer will become painfully obvious.

Stepping off the ladder was a breeze. I stepped onto an area that was already covered with decking. It was like a "big porch". This is a "piece of cake" I thought. Boy, was I in for a big surprise! That is when I encountered the "beam". At first glance I thought it will not be that big of deal. It wasn't either until got about half way across. And then there is the "rest of the story" as Paul Harvey used to say.

And now, there I was, securely anchored to the vertical steel column. I was clutching on "bulldog tight". I wasn't going anywhere. The next beam in front of me was only half the width of the one I had recently departed. No force could make me walk on that beam! What I would have given for a pair of ruby slippers. I wouldn't even have cared if I ended up in Kansas after clicking my heels. Anywhere was better than where I was at!

But alas, my uncle saved the day. He informed me that it was quite acceptable to "straddle the beam" and walk on the "bottom", commonly referred to as "crawling" which was **Lesson #2**. Well hey, you gotta crawl before you walk, right? Well, I tried it and found it to be a whole lot more palatable although I still didn't forget where I was at. It wasn't long after this discovery that I absentmindedly stumbled, literally, onto **Lesson #3**. Pay attention to where you walk! I stepped onto a section of decking without paying much attention, I might add. However, unbeknownst to me, the decking was not supported on one end. I could have set the high dive record with no water! But God or Paul was watching over me that day. I am not referring to the Apostle Paul either. I mean ironworker Paul. A man with the biggest set of biceps I had ever seen with my own two eyes and the disposition of a bobcat. As I recall, his exact words to me where, "Hey you stupid (expletive), look at what the hell you are about to do! The end of that decking is unsupported, which is okay, except you are about to die! So, get your stupid (expletive) self back over here! You didn't have to hit me in the head with a shovel! I got off that decking like a cat shot out of a cannon. Of course, I didn't go anywhere near Paul either as I was afraid he might just throw me off the building in contempt for being so ignorant.

One other item about Paul that stands out in my mind is he would light an acetylene torch in his pocket. It tends to be breezy when you are working up high so lighting anything could be a challenge. But come on………..lighting a torch in your pocket? I often wonder if he ever burnt his pants off.

Later in the morning, a young black man by the name of Henry inquired as to how long I had been an ironworker. I asked him for the time, I remember he replied it was 9:30 a.m. I then told him; "oh about two and a half hours". He looked like he was going to fall over. He told me I looked like a veteran. He asked how I could adapt to the height so easily. I was asking myself the same question inside. I thought I should win an Academy Award for this acting job. In reality, I felt like I was running through a fireworks factory with my hair on fire! Not a very pleasant feeling.

And then came **Lesson #4.** Don't drop your tools. About midday we finished on the ninth floor and were moving to the tenth. Paul (with the biceps) and my uncle shinnied up the vertical steel column to the tenth floor. My immediate reaction was, "if this is the moment where I have to prove whether I am a man or a mouse, you may call me Mickey right now". There has to be a better way. And there was the notorious right angle ladder. And since I was the one climbing the ladder I was asked to bring the tools. Okay, no problem, or so I thought. When I got to the ladder I experienced an epiphany. I was carrying two "beaters" (sledgehammers for layman) and two hook tools for dragging the sheets of decking. It required all of two hands to carry these tools. To climb the ladder required at least one hand and in my opinion, two would have been much better. One option that I briefly considered was to use my teeth. I have always prided myself on my ingenuity and this time it came in handy. The ladder was positioned next to an elevator shaft, which had a safety net draped across it. So, I very carefully deposited one hammer on the net. Ingenuous, huh? When I arrived on the tenth floor and met up with the rest of the crew, my worst fears were realized.

They could all count! They noticed the missing hammer. When they asked me as to its whereabouts, I casually mentioned that I had dropped it. And that is when Paul proceeded to explain to me in his most professional manner which contained many expletives as to why you never drop hammers when you are working in the air above other workers. I could see his logic. Worse yet, I was delegated the task of retrieving the hammer. Just before I turned into a pumpkin, one of my fellow crewmembers offered to go get it. I do not remember his name but I am eternally grateful to him.

We got all the tools in order and the rest of the day was rather anticlimactic until later when I was sitting on a beam in the middle of the stratosphere and another ironworker needed to get by me. My first thought was, one of us needs to move and as much as I would like to be accommodating, I am staying put! This particular ironworker had been around; he was far from a novice. He casually stepped over me and then stopped to chat. I did not fool this man. He knew immediately that I was a rookie; partially I am sure from the rust stains on the seat of my pants. He offered me **Lesson #5,** which is the best lesson of all and applicable to all of us. Don't look down and don't walk around with your nose up in the air!

I completed my first and what was to be my last day as an ironworker without misfortune. After giving it much thought; I concluded "ironwork is for ironworkers of whom I am not one". I am quite content to remain a groundhog. Hey don't laugh; at least I have my own holiday.

This story I dedicate to those men and women who scale the heights to construct North America's buildings, towers, and bridges. Without them, there would be no infrastructure. I also dedicate it to my relatives who are or were ironworkers out of Local 22 in Indianapolis, Indiana. My uncles Kenny and Dr. Ben Asdell and my cousin Randy Asdell. Uncle Ben is today a dentist practicing in Loogootee, Indiana. That should dispel the stereotype of construction workers "wearing a size 20 shirt and size 3 hardhat".

I read an article about the bombing of the Alfred P. Murrah Federal building in Oklahoma City, specifically about the search and rescue efforts. Ironworkers from the local in that city who built that building offered their services to the fire department to shore up the structure and make it safe for rescue efforts. After all, who better to shore up the damaged structure than the very folks who originally constructed it? At first the fire department declined their offer citing safety concerns for the volunteers but after starting rescue operations they discovered how precarious the structure really was and how badly it needed to be shored up to ensure the safety of the rescue teams and victims. So, they consented and allowed the ironworkers to do their stuff and shore up the damaged structure so rescue efforts could commence. After the World Trade Center attack, ironworkers (as well as others) were once again on the scene assisting in clean up and rescue efforts.

Sometimes I think our society looks for heroes in Washington D.C. and Hollywood when we should be looking in our own neighborhoods. Heroes include not only police and firefighters but also other rescue workers, teachers, Scout leaders, or just the elder couple next door who share their life experiences and values with our children. Heroes come in all shapes and sizes, one for every occasion.

Lunch on Skyscraper 1932 Rockefeller Center

This story was published in: "The Ironworker" trade magazine published monthly by International Ironworkers Union and the Glaziers Union Trade Magazine (2010)

Kenny Asdell

Monday, January 25, 2016

On Jan. 25, 2016, a legend passed through the gates of Heaven as Kenny Asdell departed his earthly vessel and joined the angels for all eternity leaving behind his family and friends.

Kenny was born on July 4, 1927 at the homestead of his maternal grandparents, Joseph and Cora (Blackmore) Kern, who resided just north of Scotland, Ind. His proud parents were Joseph Leland (Drib) and Sophia (Sophie) Asdell. Kenny was the firstborn but was soon followed by Drinae Fae (who passed away soon after birth), Davie Joe (Joe), Benoni (Ben), Doris June (June) and Billy Nelson (Bill).

Kenny attended Scotland High School but left early in his senior year to join the war effort by enlisting in the United States Navy, where he served honorably in WWII in the European Theater.

Upon his return from the war, he engaged in several adventurous endeavors, including building silos and grain bins throughout Michigan and Canada while employed by C&B Silo, and constructing the very first electric power lines through the Ozark Mountains of Missouri.

It was in Missouri where his brother, Joe, was severely burned in a workplace accident. Kenny was by Joe's side through his arduous recovery, even to the point of pawning his winter coat for subsistence.

Kenny loved truck driving and was an independent trucker, his first truck being a 1948 Auto Car. Kenny spent many nights away from home transporting freight and grain throughout the lower 48 states. Being a proud, 50+ year member of Ironworkers Local 22 in Indianapolis, Kenny not only worked hanging iron throughout the Midwest, he also had an opportunity to work on construction of the Alaskan pipeline in the mid-1970's.

Kenny married Vonda Fay Beasley and they had three children, Randall Kern Asdell (Randy), Marsha Dee Asdell (Dee Dee) and Sheryl Nieman (Mike). Kenny and Vonda raised their family on the homestead of Kenny's paternal grandparents, David Samuel and Clara (Stone) Asdell, which remains in the family to this day.

Kenny lived an adventurous life that most people can only dream of or experience in the movies. Even so, he always had time for family and friends and was first in line to assist those in need.

Kenny was a big man, bigger than life and he feared nothing. A fitting description of Kenny was once offered by his brother Joe: "He is bigger than a skinned elephant, meaner than a striped ass monkey, tougher than rawhide, and all man, from the ground up." This, my friend, is Kenny Asdell in a "nutshell". Kenny was a lifelong member of American Legion and VFW.

Kenny was preceded in death by his parents, Drib and Sophie Asdell; brother, Joe; and sister, June.

He is survived by Vonda; children Randy, Dee, and Sheryl; plus, many grandchildren, great-grandchildren, nieces and nephews.

A memorial visitation will be held from 2 to 4 p.m. on Saturday, Jan. 30 at Jenkins Funeral Home in Bloomfield.

Memorial Service will be held at 4 p.m. on Saturday with Mark Lester officiating.

Uncle Joe

There is Uncle Joe, he's a movin kinda slow at the junction.

This story isn't about Uncle Joe from the TV series, Petticoat Junction, in fact, far from it. The Uncle Joe featured in this article is my uncle, Joe Asdell, from Odon, Indiana. The only likeness Joe Asdell shares with Uncle Joe from the Petticoat Junction TV show is the name. Uncle Joe was born in New Castle, Indiana, in 1929, to my grandparents, Joseph (better known as Drib) and Sophia Asdell. But Uncle Joe's stay in New Castle was short lived. The Asdell family moved back home to the Scotland, Indiana, area. Legend has it that he was a pretty fair basketball player for the Scotland "Scotties". Legend also has it that he was an exuberant youth who enjoyed the social activities of the day, such as "tipping over outdoor privies" and other such non-curricular events. Uncle Joe was too young for service in World War II. His older brother, my Uncle Kenny, left school and enlisted in the Navy at 17, seeing action in the Atlantic theater. Upon his return from the Navy, Kenny found employment for Joe and himself with C&B Silo Company in Charlotte, Michigan. Kenny and Joe spent several summers constructing silos throughout the lower peninsula of Michigan. After their stint constructing silos, they found employment in Missouri running electric power lines into the Ozark Mountain foothills. It was at this time that Uncle Joe's life was to undergo a drastic change. One cold winter day, while they were trying to warm up around a wood stove in a lineman's shack, a co-worker threw some coal oil on a dwindling fire in the stove. Of course, the coal oil accelerated the combustion process and flame followed the oil to point of origin. However, the point of origin was now my Uncle Joe as the can ended up on his lap. Joe burst into flames and raced outside of the shack fully engulfed in flames (he told me that he ran out of the shack as that is where they stored their dynamite). Once the flames were extinguished, Uncle Joe was taken to the hospital. He was severely burned over a large portion of his body. The doctors never expected Uncle Joe to survive. But they underestimated the grit and determination of Joe Asdell. He did survive. His legs were so severely burned that the doctors determined they would have to amputate them. But, once again, they underestimated the grit and determination of Joe Asdell. He wouldn't allow it. Next, the doctor's determined that Joe would never walk again. But, once again, Uncle Joe proved them wrong.

Uncle Joe spent several agonizing months in a hospital bed in Missouri before being transferred to a hospital in Indianapolis, Indiana, for plastic surgery. He wasn't quite going home but he was getting closer. The Jenkins boys, Blue and Jimmy, from Jenkins Funeral Home in Bloomfield, Indiana, drove their ambulance all the way to Missouri and transported Uncle Joe back to Indiana for his plastic surgery, all free of charge. Uncle Joe was still in extreme pain and had to be kept medicated throughout the trip. On the way out of town, Uncle Joe had the Jenkins brothers stop by one of the doctors' homes so Joe could see the doctor's foxhounds. Uncle Joe was a "dyed in the wool" foxhunter in those days. Uncle Joe did recover from his burns and through sheer grit and determination, regained his ability to walk again. He started by using his wheelchair as a walker. Then as he progressed, he spent considerable amount of time in the woods, squirrel hunting. He said walking in the woods forced him to raise his legs higher and bend them further. Uncle Joe got married to Mary Ellen Crane and they had three children, Dave, Terry, and JoEllen. Joe farmed and worked as a mechanic. He also switched hounds and quarry. Uncle Joe traded in his foxhounds for coonhounds. It was about this time that I entered the picture. My Dad died before my 5th birthday leaving me fatherless and in need of a male role model. My Mom, Uncle Joe's sister, has four brothers, all who have played a significant role in my life. My Uncle Joe is the kind of man that movies are made about. To a little boy, he was a superhero. There wasn't anything my Uncle Joe couldn't do. He was rugged, strong, lean, and to use one of his own descriptive terms, "All man from the ground up". If you were with Uncle Joe, you were safe and untouchable. No, uncle Joe was intimidated by nothing, man or beast. I remember going coonhunting with Uncle Joe when I was a young boy. I didn't own any insulated undergarments so I wore two or three pairs of pants and several shirts to keep warm. I had so many clothes on that I looked like the Michelin man. I didn't move much better either. I remember following Uncle Joe through the pitch darkness with the only illumination from Uncle Joe's carbide light. I remember the shadows cast by the flickering flame dancing across the rock outcroppings in the southern Indiana hollars. I can remember Uncle Joe skinning coons by the light of the carbide flame. The flickering flames dancing across the strong rugged hands as they deftly separated the coon from his hide. As I entered my teen years, I got the idea that Uncle Joe couldn't be "kool" as he was from the "establishment". We all know that in the "60's" anyone in the "establishment" was very "unkool". But as I got older and more mature, I came to discover a wise old southern Indiana farmer and coonhunter who has forgot more about living that most people will ever know. He was an individual who had to face excruciating pain when he was little more than a kid himself to survive horrendous injuries. I purchased an old Walker hound from Uncle Joe back in 1975. He was a Finley River bred hound out of Finley River Jack. The old hound was bawl mouth all the way, even a bawl mouth tree dog. He wouldn't' bark 100 times a minute, but he would be at the tree with a coon when you got there. Uncle Joe knew what he was doing when he sold me that dog. The old dog was a "professor", he taught me everything I needed to know about coonhunting. Uncle Joe and I lived 400 miles apart and he couldn't be on hand to tutor me, so he did the next best thing. That has been over twenty-five years ago, I still have a flea-bitten Walker hound or two. I have spent a lot of time with Uncle Joe since those days, both in the woods and out. I have learned a lot about coonhounds, coonhunting, and life in general. I learned how to build up a big fire in the woods without burning the woods down (foxhunter style), I learned what wood to use to get a fire going from scratch, and I learned how to study the habits of coons. I learned how to adjust my hunting habits based upon observations of coon habits, thus giving the hounds an advantage (sometimes). I learned there is no such thing as "a good horse or good hound with bad color", especially if it is a Walker hound or Quarter horse. I learned that a man's word is his bond. I learned that integrity is difficult to earn and easy to lose. I have learned that "anyone can take the easy route". I have learned that "when the going gets tough the tough get going". I have learned that dogs as well as men gain wisdom with age. Young hot shot dogs are like young hot shot men, they lack substance. Only when their character has stood the test of time can they be declared truly great. I see old hounds advertised as "pup trainers" as to imply that they have little value anymore. These old hounds are not only "pup trainers" but they are also "boy trainers' for young men. For those who measure their success based upon nite hunt winnings and championship dogs they hunt or raise, I say you are using the wrong benchmark. How many young people have you trained? Give me an old dog, an old pickup, an old man, and a young boy or girl, now that's the making of success. I have learned that tough men with rawhide exteriors have soft interiors and big hearts. Yes, I have learned a lot from my Uncle Joe. And Uncle Joe faced another crisis a few years ago. He learned he had colon cancer. His mother, my grandmother, had colon cancer and had to undergo a colostomy. Uncle Joe decided he wasn't going to have a colostomy. Of course, the doctors argued that it might be necessary to perform a colostomy. But these doctors didn't know who they were arguing with. They underestimated the grit and determination of Uncle Joe. Uncle Joe didn't get a colostomy; once again he did it his way. Sadly, Uncle Joe lost his eldest son, Dave Asdell, a few years ago. Loss of a child by a parent is a loss from which you truly never recover. I truly believe that God sends guardian angels for us, especially when we are vulnerable and in need. I don't believe that these angels have wings and float in the cosmos. I think God uses real live people to assist and encourage others that are in despair. I think that Uncle Joe has been one of these people. He has had a positive impact on a lot of lives, especially young people who has taken coonhunting and shared with them his wit and wisdom. I owe a lot to my Uncle Joe. I thank God for the gift of such an extraordinary Uncle. God bless! Uncle Joe passed away in 2008. He fought the good fight

Diamond Jack; the "coondog professor" I got from Uncle Joe back in 1976 who taught me "the ropes".

Scotland & Cedar Springs Connection
Mark A Lester

C&B Silos at Sand Lake Grain & Elevator in Sand Lake, Michigan

If you live long enough you will most assuredly discover that life is tinged with irony and events that, on the surface, would seem not related but upon scrutiny you discover there is a connection. Sometimes it takes you a very long time before the "light goes on" and you discover the subtle connection between two nonrelated events. This "ah ha" moment recently happened to me.

Unbeknownst to virtually every Cedar Springs Bugle subscriber and Cedar Springs resident in general, there is a direct link and connection between Scotland, Indiana, a very tiny burg located in the hill country of southwestern Indiana, and Cedar Springs, Michigan, which is a small Michigan community on the west side of the state. This connection transcends three generations of Michigan and Indiana descendants.

The story begins with two U.S. Navy veterans of WW II who served together. These two become friends during their service to our country and at the end of the war teamed up to start a business venture. These two young men were John Cooper of Grand Rapids, Michigan and Ralph Baird of Scotland, Indiana. They joined together and formed C & B Silo Company, whose manufacturing facility was located in Grand Rapids, Michigan with company headquarters in Charlotte, MI. Grand Rapids was where the concrete staves (the actual walls of the cylindrical silo) were manufactured. At the time, right after WW II, Michigan was a leading dairy state and a ripe market for silos (most existing silos were of wood construction which was prone to leakage and rotting). C & B Silo needed silo erection crews to perform the silo installations. While it is, true there were large numbers of servicemen joining the workforce after the war, it was also true that the economy was booming due to pent-up demand for manufactured goods (i.e. cars, trucks, tractors, and so forth) so there was an abundance of lucrative manufacturing jobs available. There wasn't an abundance of guys in Michigan lining up to travel throughout the state and erect silos working out in the elements at the wages offered. However, in Scotland, Indiana it was a whole different story. The economy in Scotland had boomed throughout the war effort because a mile down the road was the huge Crane Naval Ammunition Depot which provided a massive amount of the munitions required by the U.S. Navy during the war. The local economy around Scotland was "drying up" at the conclusion of the war. Consequently, Ralph Baird recruited for silo erectors from the Scotland, Indiana locale. Ralph's brother, Ray, had a farm adjoining the village of Scotland providing Ralph a reliable local contact. As a result there were a large number of Scotland, Indiana residents who traveled to Michigan and erected silos for C & B Silo Company over the years. Two of these residents were my mom's older brothers (my uncle's) Kenny and Joe Asdell (Kenny being a U.S. Navy veteran of WW II himself). All these guys were very young, most being 20 and under.

There is a lot more I could add to this story if space would allow but the "meat of it is":

Uncle Joe built quite a few silos around the Hesperia, Fremont, Whitehall areas. One of the last tasks in silo construction is sealing the inside cement walls. C & B Silo used a "whitewash" mixture of lime, plaster, and cement which they brushed on while the worker was suspended on a board swing suspended by ropes from the top of the silo. Uncle Joe was performing this task in a silo up around the Whitehall area when the board he was sitting on broke in half sending him into freefall. He avoided serious injury only because he landed on the whitewash bucket which collapsed breaking his fall. This story had been Asdell family folklore for years even up to the point in time when my sister Pam married Doug Middleton, son of Rex Middleton, in 1974. Rex lived on his family's farm on 16 Mile Road south of Cedar Springs for many years. Rex Middleton had a milk route and picked up milk from farms in the same area where Uncle Joe had his accident during the same time frame. He, in fact, heard the story about the silo builder landing on the paint bucket from some of the farmers whose milk he picked up; that story became folklore amongst the farmers in the local community in the day. The time frame for this event was late 1940's; fast forward to mid-1970's when Uncle Joe visited Doug and I for what was to become our annual coonhunting rendezvous. We took Uncle Joe over to meet Rex Middleton, Doug's Dad. Doug and I never considered the fact that Uncle Joe and Rex might recognize each other; I mean after all, Uncle Joe was erecting silos on the very farms from which Rex was picking up milk. We did know that Rex wanted to meet the "man who fell and landed on the paint bucket while building a silo" who had been the talk of the town some 30 years ago. Lo and behold; their first comment to each other at that meeting was; "I know you". Doug and I were astonished. I don't think Rex knew the identity of the "guy who fell and landed on the paint bucket" until that vey instant, but he certainly had interacted with Uncle Joe some 30 years prior. It was astounding to us that these two old farmers some thirty years later would recognize each other. It was if time stood still, if only for a moment.

Thirty years later and three hundred miles apart their paths crossed again. They are both gone from this world now but there is still the "tie that binds". The third generation, Ryan Middleton, grandson of Rex and great- nephew of Joe, drives milk truck picking up milk from farms. His first route was in the Montagues area which is in the vicinity of where his grandpa picked up milk back in the 1940's and his great uncle Joe built the silos that stored feed for the dairy cows that produced the milk. When Rex picked up milk, the milk was in milk cans which held 10 gallons of milk each and he manhandled onto a truck for transport to a local dairy. When Ryan picks up milk, it is with a semi-tanker that holds about 12,000 gallons of milk that is pumped in from the farm's bulk storage tank. And I doubt any of modern farmers are still using those concrete C & B Silos such as the type built by Uncle Joe. Technology is different, people are different, but the end result is still the same and the connection is still very much in place. Truly, the world is very small. Oh yeah, just an f.y.i; Uncle Kenny's crew erected the grain bin at Sand Lake elevator which is actually C & B silos stacked up. Add to that, there is a connection to Uncle Kenny, Rex Middleton, and the old "Chicken Coop". Not that they ever squared off against each other but most assuredly had some common opponents.

Story written by Mark Lester and featured in Cedar Springs Bugle monthly periodical published in Cedar Springs, Michigan by Col. Tom Noreen. U.S. Army retired.

Canadian Black Bear Hunt
Mark A. Lester

"If I had known you could take a blow like that, I would have got you a job at the circus being shot out of a cannon" exclaimed my Uncle Joe upon my return from the hospital after being treated for injuries sustained in a motor vehicle accident in Heyden, Ontario, Canada. So began our long-anticipated bear hunting trip in Canada, which was not unfolding at all how we had planned it. My Uncle Joe Asdell and myself had made plans to join my brother-in-law, Dale, in Ontario, Canada for a spring black bear hunt. Dale was in a partnership with a Canadian outfitter in the Echo Bay area where they operated a guide service. Dale was acting as the guide for a group of black bear hunters who had reserved a spring hunt over bait. However, Dale was a houndsman who enjoyed hunting black bear with his pack of Plott Hounds and had invited us along to join in a hunt with his hounds prior to the arrival of his paying customers. My Uncle Joe was a houndsman who had owned and hunted foxhounds and coonhounds throughout his lifetime. I enjoyed following coonhounds as well, however, neither of us had ever participated in a bear hunt, either with or without hounds.

Uncle Joe had just finished planting his crops so had free time to pursue other adventures. In the first week of May, 1992, he headed to Michigan where he hooked up with me and we began our journey to Ontario to meet up with my brother-n-law Dale and his friends Al and Terry. Dale was a good friend of Leo Dollins and his pack of Plott hounds were out of Leo Dollins bloodlines. We were to hunt in the Heyden, Ontario area which is a short distance from Sault Ste. Marie, Ontario. Once we arrived in Heyden, we stopped to purchase a bear license prior to heading for Dale's bear camp located out in the "bush". After purchasing our license, we were in the process of leaving the sporting goods store in my Ford Ranger pickup waiting for traffic to clear on Highway 17. As we were waiting, a car heading our direction at a high rate of speed veered onto the gravel shoulder. At first, I thought they were turning into the parking lot of the store we were exiting but there was no turn signal activated that would indicate their intentions plus they were not slowing down, rather they were heading right at me. I shifted the truck into reverse and started backing up to avoid a collision but I was not fast enough and their vehicle hit me square on the driver's side front tire wiping out the front clip while violently spinning the truck around. Stupid me was not wearing a seat belt so my forehead met the windshield, my chest slammed the steering wheel and my shoulder relocated the steering column. It was a heck of a hit. The good news; the local volunteer fire department was practicing next door so first responders were on the scene immediately after the accident and luckily, I had backed the truck up avoiding a hit to the driver's door which would have resulted in tragedy. I received firsthand experience with the Canadian health care system, which turned out fine by the way, and had no broken bones, just massive bruising including bruised ribs, a separated left shoulder and two "goose eggs" on my forehead. Of course, I also had a totaled-out vehicle in a foreign country which became a challenge but not immediately. Fortunately, my Uncle Joe came through the whole event unscathed. As I was to discover, the driver that hit me was drunk.

So began our long-anticipated bear hunting expedition. After being checked out at a Sault Ste. Marie, Ontario hospital, I hitched a ride from the hospital back to the accident scene from an Ontario Provincial Police Officer. He was a very nice guy and we shared a mutual acquaintance, I knew one of the self-defense instructors for the Ontario Provincial Police from my karate training. In my absence my Uncle Joe had made friends with the local teenage boys. I think they found him to be a "novelty". A personable "down home" old Indiana farmer with a southern dialect who made friends easily. The local boys were able to get word to our compadres at the bear hunting camp and upon my return I found the guys unloading our gear from the damaged truck.

The guys had a camp set up on Provincial Land next to a creek. Dale had purchased a vinyl clad cabin which was designed to be "broken down" and transported so "roughing it" was not all that rough, which worked out great for me as I was in "rough" physical condition but I was alive and mobile. We even had bunk beds to sleep on.

Neither I nor my Uncle Joe had ever been on a bear hunt, with or without dogs and Uncle Joe had never been to Canada so this adventure was a total new experience for us both. When we packed for the trip, Uncle Joe and I packed for cool weather but upon arriving in Canada we found the daytime temperatures running in the 70's (Fahrenheit of course). Ironically, Dale told us when he arrived a couple of weeks prior, he was pushing snow with his truck bumper. There was just a little snow left in the woods when we arrived. We soon found that we were "overdressed".

Having never bear hunted before, I had no prior experiences to use as a reference and comparison to our Canadian adventure. Dale did tell us that spring bear hunting was not analogous to fall hunting. When bears arise from their long winter slumber they are lean and in search of fiber to restore functionality in their digestive systems. The challenge we faced was these bears would run long distances and be reluctant to climb a tree. As we were soon to learn, Dale wasn't exaggerating. Our primary hunting area was around Searchmont area. There is a large ski resort located in Searchmont, which provides you a clue as to the terrain we were hunting. It wouldn't be considered as mountainous but it was not far from it. The Goulais River runs through this area, in fact, the hounds ran a bear across the river once. The Algoma Central Railway, famous for the fall color and winter train tours through Agawa Canyon Park runs through the area where we hunted. Dale shared with us that they had treed a bear along the railroad grade a few years prior. The largest impediment we encountered was lack of roads. We traveled logging roads mostly, some active and some abandoned which were left over from prior logging operations. We would rig the hounds up these logging roads to strike a bear. Once scent was detected and we were made aware by the bawl of the hounds, we would verify by looking for sign, which would be bear tracks in the dust or bear scat. Dale told me that he had seen hounds run a moose on occasion which I found interesting. I have seen hounds run amuck, but never run a moose. All kidding aside, I was surprised a moose would even run from a dog or that a moose would run at all due to their huge stature. We did encounter some moose sign, tracks as well as scat, but never saw an actual moose. Our biggest challenge was getting to the hounds. Once they struck a track, the bear would head out of the country and were soon out of hearing. Most of the time we were guided by signals from the tracking systems (this was prior to GPS). Often the hounds would travel outside the range of the tracking system in which case we used "best guess locating system".

Since the logging roads were not necessarily interconnected, many times we would "backtrack" to a main road to find logging roads in the direction our hounds were heading. This was repeated several times during our bear hunting experience. The hounds might run for 7 or 8 miles while we were forced to drive 25 to 30 miles to reach them due to lack of roads. I do not remember how many days were hunted but we never did see a bear in a tree. The only time I recall them treeing was a tree that extended up to a rock ledge. It appeared the bear climbed the tree and exited onto the rock ledge. I do not remember if the hounds picked up the track on the ledge or not. The hounds lost one track in the Gouglais River.

On one trip down a logging road, the pickup I was riding in dropped through a culvert. The culvert collapsed and the front wheels dropped through. That was quite a jolt, especially for my sore body. The guys jacked the truck up with a high lift jack and winched the truck out and we were soon on our way. We had to backtrack a bit but we were fairly well accustomed to adjusting our routes. Anybody contemplating purchasing a new pickup and wishing to find the toughest "pickups" manufactured should consult with a big game houndsmen and see what pickup brand they use.

Every morning Dale would check on the bait piles he had set out for his soon to arrive "paying customers". Bears were visiting at least some of these sites. We never turned the hounds out at any of the sites.

A day or two after the accident, Uncle Joe and I borrowed Dale's truck and stopped at a nearby campground where we could get a shower for a fee and use the payphone to call home. We had to let everyone know about the accident and reassure them we were both all right plus I needed to contact my insurance company and explore my options for a rental vehicle. I was advised by office staff at my insurance agent's office to cross the border and rent a car in the U.S. So, based upon that information, Uncle Joe and I headed for Sault Ste. Marie, Michigan. Dale's truck had a 100-gallon auxiliary fuel tank and there was a .22 caliber rifle behind the seat, which, in hindsight I should have removed. When we reached the border, I explained our situation to the Canadian border guard, including the presence of the rifle and the oversized fuel tank, and told him we were on a mission to acquire a vehicle and would be back shortly. All seemed to be well. As we soon discovered, there were no car rental agencies in Sault Ste. Marie that I could use to rent a vehicle through my auto insurance. So, we gave up our search and headed back to bear camp. However, my injured shoulders were aggravated from steering the truck so Uncle Joe took over driving. Upon reaching the border, we were greeted by a different border guard who didn't know of our plight and didn't care. She asked Uncle Joe if we had any weapons. His response was "yes". She then asks what kind. His response was "a gun". She then asked what kind of gun and his response was "a rifle". She then got aggravated and told him to "pull it over". We were going to get searched. I watched all this unfold thinking "is he messing with her on purpose or is he having trouble hearing her or what, because, as anyone who has ever passed through border security can attest, they have little patience and if they feel you are being evasive, they are not hesitant about having your vehicle and/or your person searched. And this all happened prior to 911.

The vehicle we were driving was not ours which generated questions from the border patrol but even worse, I didn't know if there might be something in it that could cause concern, aside from the gun which we had already declared. The search was going fine until the border guard searched our clothes bag containing the dirty clothes from our shower that morning. The border guard ran his hand inside the bag of dirty clothes looking for contraband whereby, Uncle Joe observing this retorted "If I had known you were going to do that, I would have put a dog turd in there". I was like "wow", now what. The border guard, however, did not respond, he just finished his search and sent us on our way. Oh, Uncle Joe, I cannot help but love you. Uncle Joe did develop a fan club. He was a celebrity with the young folks in town whom he had met after the accident. When we drove through town if any of them saw him, they would wave and shout out to him. We stopped at a restaurant for lunch and some of them followed us inside so they could hear some more Uncle Joe stories. The locals all knew what we were up to. Dale was not a stranger having hunted in that locale for several years plus they recognized the dog boxes. The residents I talked to were supportive of bear hunting, they considered bears to be a nuisance so were in total favor of population control. The bears would break into their remote fishing cabins and destroy the contents which did not win them any friends.

Since I was never able to procure transportation, I had made contact with a very good friend to transport us back home. He was to meet us in Sault Ste. Marie, Michigan on our last day. The morning of the last day, we decided to have one last hunt in hopes of getting a bear up a tree. But of course, as is often the case, the hunt didn't go as planned. The hounds did hit a bear track and became scattered so we divided up into two groups. Uncle Joe, Dale, and Terry were in one group and they ended up walking about 7 miles through the bush. Uncle Joe made a believer out of Dale and Terry, as they would need to stop and get their breath but he would keep going and with him being 30 years older than either of them. They told me later "he is one tough guy". I was with Al and we became separated causing me to walk about 4 miles back to our bear camp. As a result, we were late getting to our rendezvous point with our ride home and had no way to contact my friend to let him know we were running late but we still coming. But in the end, albeit late, we were able to get across the border and meet my friend, who fortunately for us, was still waiting there. Never did tree a bear though, which was disappointing and even more so for my Uncle Joe.

We met up with out ride and left Sault Ste. Marie, Michigan, heading south thus ending our bear hunting adventure which definitely didn't pan out as we had anticipated. Uncle Joe went back to his southern Indiana farm. I worked with my insurance company getting a settlement on my totaled out pickup. It actually went well, the only hold up was the insurance adjuster wouldn't travel into Canada to inspect the vehicle so the insurance company had it towed back into the states to appease him. That delayed the payoff for a few days. Plus, I went through a couple months of physical therapy to restore functionality in my shoulders.

At the end of the day, we were fortunate not to have been more seriously injured in the accident. We did have an "eye opening" experience bear hunting with hounds in some very rough terrain. Uncle Joe was able to visit Canada for the first time. Uncle Joe gave up smoking while we were in Canada. He was a three pack a day cigarette smoker and was wearing a nicotine patch while we were there but still smoking as well. However, when he ran out of cigarettes while in Canada and discovered they were $7.00 a pack for cigarette brands he didn't recognize, that did it. He quit for good. This bear hunt occurred almost 30 years ago. We never went back and tried it again. Dale and my sister are no longer married and he resides in the Upper Peninsula of Michigan. Al lives in the Sand Lake, Michigan area. Terry moved to Colorado and bought the Kokopelli Ranch in Bellvue, Colorado where he raises Yaks, and Uncle Joe passed away almost 13 years ago. And, as a side note, Ontario no longer has a spring bear hunting season. Dales' paying customers had a significantly higher success rate hunting over baits than we did using hounds. This should dispel the justification anti's espouse for banning hunting with hounds as (says them) it gives the hunter and the hounds an unfair advantage. But we all knew that was a false assumption anyway. Happy Trails!

Ramblings of an Old Man
The Curse
Mark Lester

I had a "déjà vu" night this past week. In a moment of shear weakness and empathy I loaded up my thirteen-year-old Yadkin River bred female and took her to the woods. Taking her hunting has become a rarity as "she is deaf as a stump", sleeps most of the time, and suffering dementia issues. Funny thing, the last statement pretty much describes me. My wife is adamant that I have no business going to the woods either, and for sure not by myself. Like I said, in a weak moment I loaded her up and away we went. She has never shown any hesitation about hitting the brush but she is negligent about checking back in. Back in the day, when we were both younger and had our wits about us, that did not present much of an issue. That is not true anymore though. Our night started well, she struck and treed and looked good doing it. Circumstances deteriorated quickly from there on out. After heading her out from the tree she did not resurface for several hours. As I waited in the truck for her to either tree or come in, I had plenty of time to reflect and that

was when the "light came on in my head" and I decided I was "cursed" by my Uncle Joe who first introduced me to coonhunting. The more I thought about it, the more convinced I became that I most likely would have been farther ahead to have been blessed with a favorite Uncle who was a croquet aficionado or into badminton or pinochle or something, anything but coonhunting. The next morning, I called my brother-in-law, Doug Middleton, who has been my coonhunting partner and best friend for 50 years or better and explained my recent epiphany to him. The longer we dug up memories the more convinced we both became that Uncle Joe had led us astray.

Uncle Joe Asdell   2006

As if introducing us to the "sport" of coonhunting wasn't enough of a curse, we were indoctrinated by one tough guy who was serious about his hounds, his hunting, and not intimidated by man, beast, weather, or terrain. There was no hill high enough, no holler deep enough, no swamp mucky enough, no woods thick enough, no briars sharp enough, no man belligerent enough, nor any hound independent enough to cause him to retreat. I told Doug, you know those "so-called deer trails" in Michigan and Indiana were not made by deer, nope, they were made by our "butts" dragging on the ground trying to keep up with that guy, i.e. Uncle Joe. Joe's first hounds were fox hounds so he was always partial to running dogs. He desired hounds that had the ability to move a track. He would not feed a hound for very long that constantly got "bogged down" on tracks and spent the night "boo'ing around" stomping down the fauna in the same area. If a hound was an "outlaw" even better for Uncle Joe. Running trash was not a "cardinal sin" to him a philosophy he passed along to Doug and me. Hounds generally run trash for the excitement and the "fast action" which means they have "desire" and generally they get quite proficient at track running. You get a trash running hound broke you will have yourself a coondog that can make a coon grunt and seriously look for a tree to climb. With that being said, I am not sure that running the "hair off a coon" matters anymore, as I don't recall when I saw the last good cornfield race. Uncle Joe had a female hound back in the 1970's he called Maggie who was just like him. She did not know the meaning of quit. She would drift a cold track and might not open at all until she had the track figured out well ahead of other hounds that might be opening on the same track. She broke every rule in the book to tree a coon. She didn't quit, wouldn't come in until she was good and ready, usually at daylight. Old Mag was created for my Uncle Joe. She would go as far and as hard as necessary to tree a coon. She was out of Gann's Finisher on the top side and House's Hawk on the bottom. She was a black & white open spotted Walker which is how my uncle liked them although I have heard him say, more that once, "there is no such thing as a good hound with bad color". One of my uncle's good hunting buddies was Virgil Flinn, who along with Joe was a member of the Elnora Outdoor Club. Virg was big into competition hunting and, as a result of attending hunts made a lot of hunting friends in various parts of the country. Virg brought Uncle Joe to Michigan before Joe got us into the coonhunting game and they hunted with guys Virg had competed against. One of those guys was Clifford Parks from Fulton, Michigan who campaigned a Redbone he called Cherokee Red. Parks was so enthralled with Mag that he wanted to breed his Cherokee Red dog to her. Parks never got my Uncle convinced to do it however.

At that time, Virgil had a Walker hound he called Shiloh Duke. Duke was out of Miller's Rock on the topside and Miller's Little Joker on the bottom side. Duke was a black & white open spotted Walker as well.

Joe did breed Mag to Duke and from that breeding, Doug and I each obtained a pup. Both pups had a ½ white face but on opposite sides. Doug's pup was a female he named Belle and mine was a male I called Joe. Those two pups grew up to be a lot like their mother, Mag. They were never idle. I saw Joe tree his first coon at 6 months old and run a deer the very next night in the very same woods. That was a precursor of what was to come. Those dogs were "outlaws" pure and simple. Most guys would probably have given up and put them down but we had Uncle Joe coaching us. We would get discouraged and Uncle Joe would give us the "pep talk". Don't give up boys, you get those two hounds broke and you will have the two best coonhounds in Michigan. They did tree coons but man oh man, if it was a slow night you better look out. There were plenty of deer around so if the coon weren't moving the deer would be because those two outlaws would move them. And, boy oh boy, they could run a deer and do it right. They could have a deer moved out of the section before most guys would have a clue as to what was going on. Finally, after several lectures, Uncle Joe got through to us to quit hunting those two together. Belle came around pretty quick after that and started bothering the deer less and less. Joe, though, he struggled a bit. I remember a night where we hunted from dusk to dawn and in that time frame, Joe ran 7 separate deer tracks and treed 4 coons. I had a shock collar on him so could end the deer chase rather quickly which made them all fairly short lived. Joe, the hound, got killed in an accident when he was 4 years old, and no, I did not do it. The last season I hunted him he ran 7 deer the entire season so he was definitely coming around. I had tried all the "tricks" such as putting him in a pen with a buck deer. At that time there was a guy by the name of Ben Hart over by Laingsburg, Michigan, who had a buck deer he used for breaking dogs. I hauled Joe over there and left him for a week. The very first night I turned him loose after getting him home he ran a deer. He must have become buddies with Ben's buck deer while he was there.

High Scoring Dog of the hunt at English Days 1975 U.E.B.A.F. was Grand Nite Champion Shiloh Duke, owned by Virgil Flinn, Oden, Indiana.

I tried the other methods that some guys would swear by but none made any difference except for using the shock collar when he was in the act of running a deer but that was not a one-time cure, it was a long process. He was actually frightened of a deer hide or deer body part. Uncle Joe explained it, that dog had no use or desire for a deer, he was in it for the chase. It was an easy track to run and he could "put his head up" and run like the wind. I actually witnessed him run a coon out of a corn field like that once. That Joe dog was running that coon so hard he could hardly bark, he was literally grunting. He left the other dogs in his dust cloud. That coon had to grab the very first tree he came too or Joe would have caught him. Like I said, Joe the hound could "hands down" run a track and for an old foxhunter like my Uncle Joe, that was a "heavenly" sight. Hind sight being 20/20, I should have probably tried him on bear.

Joe and Belle as puppies. They looked like "bookends".

Back in those days, Uncle Joe would make a trip to Michigan every fall to hunt with Doug and me if he got his corn shelled early enough and he generally was able to do that. He said he needed to come up and get us straightened out and he was generally stating fact. Doug and I would plan to be off work when Uncle Joe was up but we were not always successful. As coonhunters can attest, hunting all night and working all day can render one in an irritable mood, ha ha. Hunting with Joe was always an adventure. In the early days, Doug and I both lived in Cedar Springs, which is north of Grand Rapids, Michigan but I later changed jobs and moved south of Grand Rapids. Joe and I would travel back and forth between my house and Doug's to hunt. One night as we were returning home, Uncle Joe wanted to stop for breakfast. About 3:00 a.m. or so, we were coming through Grand Rapids and so we stopped at Denny's Restaurant for breakfast. We were dressed in our Carhart overalls with hip boots. I had skinned a few coons so there was blood on my boots and overalls. Joe almost never went anywhere without a cap so he removed his Wheat light and wore the hard hat for his light into the restaurant. We made quite a spectacle when we walked through the door. It got pretty quiet inside the restaurant for a while. I think the bar crowd was in there and I think there might have been some concern about what mountain we climbed down from or what psycho ward we might have escaped from. No one hassled us so we enjoyed a peaceful breakfast.

Joe when he was about 2 years old

There was the night Joe built us a fire in the woods using pocket lint and sassafras leaves. It was getting up close to daylight and there was a heavy frost on the ground, the dogs were trying to work out a track so Joe built us a fire just like they did back in his fox hunting days and we sat back and enjoyed the warmth, the dogs trailing, and his tales. Those were definitely the "good old days".

Mark Lester (myself) and Uncle Joe Asdell   1986

As I sat in my truck waiting for my old dog to come in, all these thoughts were being resurrected from the archives of my brain which was comforting on one hand, while surrounding me with an air of melancholy. I am thinking my Uncle Joe left Doug and me with a curse or, maybe, an addiction. Here we are, still messing with hounds and trying to tree an unlucky crippled raccoon for what reason we probably could never answer. But you know what? Through it all, I would not give up the memories I have just shared with you or those I haven't for any amount of money or other compensation. Maybe Doug and I don't walk as fast as we once did, nor as far, but we are still walking. We might trip and fall once in a while but we always manage to get back up and continue on our journey. We can reminisce about our many hunts with Uncle Joe and chuckle at his homespun humor and witticisms while we contemplate how to get to a treed hound that is miles away with the Grand Canyon and three rivers separating us. But you know what? It has surely been a good ride!

Ramblings of an Old Man
"Drib"
Mark Lester

Pictured here is "Drib", a model 580 Remington single shot .22 caliber rifle. I have carried "Drib" all over the state of Michigan and shot more coons with it than I can possibly remember. A coonhunting buddy told me that certain model 580 Remington's are collectors' items and worth a fair amount of money. This particular Remington does not fall into that category but it is still worth a fortune to me.

Pictured here is another "Drib". This "Drib" is my maternal grandfather and standing next to him is my grandmother Sophie. Drib was born Joseph Leland Asdell in 1906 in the Scotland, Indiana area. He acquired the nickname of "Drib" while playing in a high school basketball game where he entertained the spectators with a display of his dribbling proficiency. This was at a time when basketball was primarily a passing game. He was given the nickname of "Drib", shortened from Dribble, and it stuck with him from 1920 until he died in 1990 and, even today, as it is engraved on his headstone in the Scotland Cemetery. The rifle, "Drib", I acquired from him in the mid-1970's. He had bought it and kept it in the trunk of the car after he learned that my Grandmother had purchased a .22 caliber rifle for him as a birthday present. He didn't want her to know that he bought a rifle so he offered it to me. My Grandmother went to her grave without ever having acquired any knowledge of the "covert" rifle transaction.

My Grandfather was quite a man. He was a patient, kind, level headed individual who was well liked and well respected. When I was a teenager, I accompanied him to a strawberry patch owned by a lifelong friend of his by the name of Eddie Hostetler. When we were done picking berries, he told Eddie that I was "June's boy", Mark (my mother was his only daughter). Eddie said "Drib, you are a dwarf, all your kids and grandkids tower over you" and then turning to me he added "son, every once in a while, you need to get down on your knees and look up to this man as he has earned it". There was no argument from me, I already knew that. My grandfather had been a school teacher during the Depression, which was a very tough occupation for someone trying to support a family. It had never dawned on me that he had attended college until I tried to impress him with my ability, or lack thereof, to speak French. I studied French as a freshman in high school, so I attempted to impress him with my French speaking proficiency by asking him "Parlez-vous francais?" (do you speak French?) to which he replied "Oui, voulez-vous un café?" (yes, would you like coffee?). I was stunned. How could that little old man in the bib overalls speak French? It was then, when I was in the ninth grade that I learned that my Grandfather had attended two years of college and obtained a teaching certificate, which was the educational requirement for teaching at the time. And then there was the time a few years later I was trying to impress him with my harmonica playing expertise when he said "let me see that thing" and proceeded to belt out a "chicken reel" which he called "dog on a fence post". That was the day I learned he had, in his younger days, played a fiddle, guitar, banjo, piano, and of course, the harmonica. There you go, as they say, never judge a book by its cover.

Papaw Drib was an old time coonhunter. He always kept a hound around to hunt and he was not particular about breed or papers. He would take my mom and my uncle's hunting when they were kids usually with very limited success. My Uncle Joe, the "diehard coonhunter" would tell me "Daddy never had a dog worth 15 cents, and I would swear off hunting with him every time we went until the next time, when he would convince me once again to go along". In all fairness, "back in the day" the coon population in that area was very sparse, just the sight of a coon track would make the local newspaper.

As most of us surely know, it is really hard to "make a dog" if there is no game. He told me once that he had skinned far more possums than he ever did coons.

One of my Grandpa's talents that I most enjoyed was his story telling ability. He could keep you on the edge of your seat as he weaved his way through a good tale. Storytelling, in this modern age of "social media" has become a "lost art", which is a really sad loss for our younger generation.

Many years ago, my Grandpa had a good hunting buddy by the name of Van Feutz. Some of my favorite stories revolve around the exploits of Van and himself. I will share a couple of those exploits.

There was a local Conservation Officer that Van apparently had history with, an unpleasant history at that. As a consequence, Van enjoyed getting under the "officer's skin". Van was rabbit hunting in the river bottom around Newberry, Indiana, when the afore mentioned Conservation Officer drove up. Van was standing in a picked cornfield letting his Beagles loose when the Conservation Officer asked him if he was letting his dogs loose or gathering them up. Van replied "well I am letting some loose and gathering some up". The CO told Van he wanted to see his hunting license and asked him to come out of the muddy cornfield. Van told him, "I am busy, you need to come over here". The CO responded that he didn't have any boots to which Van replied, "even a fool knows enough to bring his boots when he is hunting". I do not remember the outcome of who walked to whom. However, my Grandpa told me that Van must have carried every hunting license he ever purchased in his wallet. He said Van would pull out a license and hand it to the CO who would look at it and tell Van it was expired after which Van would pull out another one only to get the same response from the CO. Papaw Drib said that Van must have repeated that process at least a dozen times before he finally produced a valid license. It was my Grandpa's opinion that the CO thought he had Van and could nail him for hunting without a license only to be met with disappointment.

Papaw Drib told me that he and Van were sitting in their car one night when their favorite CO pulled up next to them. They were just about to get out of the car and turn their hounds loose. The CO asked them to produce their hunting licenses to which Van replied, "we aren't in the act of hunting so we do not need a hunting license and you don't need to see it anyway". There was a short verbal sparring match between the CO and Van before the CO finally relented. Van and my Grandpa decided they better move on to somewhere else and as they were preparing to drive away the CO flagged them down. Seems his car wouldn't start and he was seeking a ride to town. Van told him "walking isn't crowded" as he drove away. Van and my Grandpa did stop at the diner in town and tell the owner there was a stranded Conservation Officer north of town and he might want to call someone to pick him up. Papaw Drib told me the CO was driving a 1937 Pontiac which will give you an idea of the timeframe for this event. He also told me that neither he nor Van had a valid hunting license that night.

There is a story I heard my Grandpa tell many times about an event featuring a Walker hound my Uncle Joe had back in the 1950's that he called Rowdy. I don't know if old Rowdy was registered or what he was out of but he was apparently a really good hound and developed a reputation as such in the area. Rowdy's reputation generated attention from other hunters who had an interest in going hunting with him. On one particular hunt, some other hunters from out of the area participated with their hounds. Rowdy commenced to tree and was joined by the other hounds. When the other hounds joined him at the tree, Rowdy pulled and trailed up farther up the holler where he treed a second time. Rowdy repeated this process three more times before settling on the fifth tree where he remained until the hunters shot out a big sow coon. Rowdy had managed to tree the litter individually on five separate trees making it appear as if he knew where each coon was at before ever striking their track. Papaw Drib did say that Uncle Joe was less than happy when Rowdy started pulling trees. Papaw said when they got to the last tree, old Rowdy was treeing his heart out with his tail wagging as if to say "lookee here at what I done. Treed five coons in a row". Top that why don't you". Funny thing, after my Papaw Drib had passed away, I was chatting with a longtime friend of his, Russell "Tussy" Summerville, who shared a story eerily similar to the Rowdy story. In Tussy's story he was hunting with Uncle Joe and a blind Black & Tan hound Joe had he called Bud. Joe had old Bud in the early 60's and Bud was indeed blind; he became blind after a bout of distemper but Joe still hunted him and Bud learned to adapt to his condition quite well. In Tussy's story, Bud treed five coons in a row, just like Rowdy had but there were no spectators that time, just Uncle Joe and Tussy. I do remember Tussy exclaiming, as he held up five fingers, "5 coons with a blind coondog". The two stories are remarkably similar and it seems astounding that this could happen twice but sometimes "the truth is stranger than fiction".

Here is a picture of Papaw Drib and me at his 84th birthday party in 1990. It was not long after this picture was taken that he went to his eternal reward. So, retracing my footsteps back to the beginning of this story, I still carry that old Remington single shot rifle I acquired from my Grandpa which, in his memory, I fondly refer to as "Drib". That rifle has similar characteristics to its namesake; it is simple, it is dependable, it is rugged, and it makes me better. My Grandpa Drib made me a better person, that old single shot rifle makes me a better shooter.

Final Thoughts

This has been a challenging endeavor for sure, similar to threading a needle. What to include, how to share without "offending sensibilities", and who to include. I am fairly certain I did not hit the bullseye 100% but feel that I at least hit the target. This final chapter is my last chance to tidy up and insert some stories that I skipped over but believe need to be included in the final narrative. So, let's begin.

Violators

Kenny, as well as Joe, killed a lot of whitetail deer in their lifetime. However, I don't know that either of them ever purchased a license. Nor do I know how many they ever killed in the daytime or during season. Before you get your hackles up, let me explain something. The state of Indiana did not have an open deer hunting season prior to 1951. Also, Kenny & Joe lived next to the Crane Naval Ammunition Depot (which today is known as the Naval Surface Warfare Center), which encompasses over 62,000 acres, a lot of which is forest. Deer hunting was not allowed on this Naval property until 1961. Needless to say, without predators (natural or otherwise) the deer population exploded on the base spilling out into the surrounding area. This abundance of hungry whitetail deer who viewed crops as their personal "cafeteria" much to the chagrin of the farmers whose profits were suffering. Farmers wanted to see the deer gone and guys like my uncles, Kenny and Joe, were happy to assist, which they did. For several years venison was the main table fare at their homes. They would sneak into the base and harvest a deer or two and then sneak them back out. Joe had, at one time, worked for the Forestry Division on the base so he knew the territory quite well. Kenny's wife, Vonda, worked on the base at the credit union so she had a pass. Kenny told a story whereby he killed a couple of deer and had made arrangements with Vonda for her to pick him up. When she arrived, he loaded the deer in back of their station wagon and covered them with horse blankets. As Vonda was exiting the base through the gate one of the deer raised his head up, apparently not dead. Luckily for them, the base security guard did not notice it.

My Uncle Bill told me about hunting with Kenny on the base. He said they saw a buck across a holler who was so far away he looked to be about the same size as the fingernail on your "pinkie finger". Kenny told Bill, watch this, I will drop him with one shot. Bill said "yeah, sure you will". Kenny did have a scope on his rifle and just like he predicted, when he touched off the rifle the deer dropped where he stood. Bill said 'I couldn't believe it; I still can't believe it and I watched it". Bill said they killed three deer during that hunt so Kenny found a pole to which they tied the deer. Bill said Kenny picked his end of the pole up with no visible effort. Bill said it was all I could do to pick up my end, let alone carry it. He said it was getting daylight so they couldn't waste any time getting off the base. Bill said he managed to stumble along behind Kenny headed to the perimeter fence. He said once they reached the fence Kenny hoisted his end of the pole over with no difficulty. Bill said that he couldn't lift his end high enough to get it over so Kenny had to help him. In all fairness, Bill would have been pretty young at the time.

My grandpa Drib told me about Joe and Bill coming to his house in the middle of the night asking him to help them pick up a deer they had just shot. They said they thought someone might have seen them so they didn't want to take their vehicle back to that area. They wanted Drib to take his car and they would load the deer into his trunk and bring it back to his house where they could process it. My grandpa wasn't too keen on the idea but he did acquiesce. He said Joe and Bill snuck the deer up to his car and loaded it in his trunk. He said he was on "pins and needles" all the way home. They hauled the deer down to his basement and processed it there. He told them that he would never do that again so don't ask.

Papaw Drib also told me that while Kenny was "trucking", Joe and Bill had deer hanging in a van body semi-trailer Kenny had parked in "downtown" Scotland. Kenny knew nothing about it. Could have been the one-time Kenny might have gotten "nabbed by the law" when he was totally innocent. But he didn't.

The "free-lance" deer hunting didn't last long after the Navy opened up the base to hunting. There is a lesson here for those folks that naively believe the "banning of hunting would ban hunting".

The ban would eliminate the state from collecting revenue from hunting license sales but it would not eliminate hunting, not from diehard hunters anyway. Just a side note, Kenny did harvest a mule deer out west, Colorado, I think. He did that totally legally, may have been the only legal deer he ever killed, I don't know. I was told that he killed that deer at 1,000 feet. That's some pretty good shooting. Some people may find irony in the fact that Kenny killed so many deer yet had so many deer as pets. I, personally, do not find any irony in that at all as I have killed a large number of raccoons in my life but yet have had many pet raccoons. I think that after you pursue an animal you learn about it, you gain respect for it, and admiration. Maybe that's "twisted, I don't know.

Speaking of raccoon hunting, Uncle Joe was riding in my pickup as we were returning from hunting with my brother-in-law, Doug, very late at night. I was driving and Joe was asleep. I was fighting to stay awake so I turned off the heat in the truck and when that didn't work, I rolled the window down. Joe woke up and yelled; "who turned off the heat". I said "I did, I was trying to keep from falling asleep". Joe said "the hell with it, turn it back on. If we are going to die let's die warm".

My brother-in-law Doug, did nod off while driving back home from hunting with me and slid along a guardrail. Uncle Bill was asleep in his truck when that happened. The only injuries were to the truck and the guardrail.

Most of the ground my Uncle Joe farmed was owned by other people which he farmed on a "sharecropper" arrangement. Joe and the landowner would share the cost of putting in the crop and any income from the harvest on a 50/50 split. Plus, the landowner would pay a fee for the harvesting which Joe said stemmed from the old days when harvesting crews would travel from farm to farm. Most of the farms he picked up after the owner retired from farming and they generally knew Joe and trusted him. He had a farm that was owned by a widow whose husband had chosen Joe to farm his land after this man quit farming. After this lady's husband passed, she continued to keep the land and have Joe farm it. Joe had farmed it for about as long as he had been farming.

Joe had gained a lot of respect for that lady. Joe told me several times he would not quit farming as long as Margaret Richardson was alive and still owned her farm. He was afraid someone else might take advantage of her. And he didn't. Joe told me about walking fields looking at sparsely germinating corn during a drought trying to decide if it would be more profitable to ride it out and be satisfied with whatever crop he might get or gamble and replant hoping for more moisture. He said he agonized for days. One of the landowners told him "Joe you need to quit worrying". Joe responded, "if I didn't worry at least a little bit, you wouldn't want me farming your farm". And that's how he was; tough as rawhide on the outside, but inside, he did care.

I have already written about Kenny and some situations where he had helped others. I didn't share about the guy working on the water tower who suffered a heart attack. The local fire department either couldn't figure out how to rescue him or were afraid to climb the tower to where the guy was at. One of them suggested they get ahold of Kenny who just happened to be home. Kenny did come to the scene and with his assistance, the firefighters and paramedics rescued the stricken man.

Or that Kenny made a habit of calling others, mostly other seniors, on a regular basis. One of these phone call recipients was my grandpa Drib's cousin. Kenny called her every morning and one morning she didn't answer. He tried several times with no success so he decided he better check up on her. He found her on the floor, she had suffered a stroke and couldn't get up. Had he not taken the initiative to check on her she very well may have died.

There you have it. They don't make guys like this every day. One of Uncle Joe's sayings was: "all man from the ground up". I think he described himself. Or I have heard him say, sometimes in referencing Kenny; "bigger than a skinned elephant and meaner than a striped ass monkey".

Whenever you were ready to give up and throw in the towel due to frustration and begin spouting off: "I can't do this", my grandpa Drib would respond; "Can't never did anything". Amen brother, isn't that the truth. If you say you "can't do something", you will be right 100% of the time. That, my friend, is a "self-fulfilling prophecy".

In his later years, Joe would say; "it doesn't cost anymore to think positive than it does to think negative". And that is so very true. You pick your attitude. Or your nose, your choice, I guess.

Last thing I wish to leave with you: my Uncle Joe told me once that "he always felt he was the closest thing I (meaning me) ever had to a dad". I have always felt the same way. I have always looked up to him, as a father figure, mentor, and guide. I have always looked up to all my uncle's and my grandpa in similar ways. I suppose at the end of the day, the most important thing is that "I have always looked up to them".

The sun always rises after the storm.

Mark A Lester

Made in the USA
Middletown, DE
10 November 2021